JENNIE HUIZENGA MEMORIAL LIBRARY

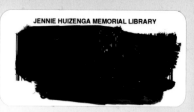

HOLT Science
Spectrum A BALANCED APPROACH

Datasheets

This book was printed with acid-free recycled content paper, containing 10% POSTCONSUMER WASTE.

HOLT, RINEHART AND WINSTON

A Harcourt Classroom Education Company

Austin • New York • Orlando • Atlanta • San Francisco • Boston • Dallas • Toronto • London

Holt Science Spectrum: A Balanced Approach

Datasheets

Copyright © by Holt, Rinehart and Winston

All rights reserved. No part of this publication may be reproduced or transmitted in any form or by any means, electronic or mechanical, including photocopy, recording, or any information storage and retrieval system, without permission in writing from the publisher.

Teachers using HOLT SCIENCE SPECTRUM may photocopy complete pages in sufficient quantities for classroom use only and not for resale.

Printed in the United States of America

ISBN 0-03-055573-6

1 2 3 4 5   862   04   03   02   01   00

CONTENTS

Datasheets

Copyright © by Holt, Rinehart and Winston. All rights reserved.

CONTENTS, CONTINUED

Copyright © by Holt, Rinehart and Winston. All rights reserved.

CONTENTS, CONTINUED

Copyright © by Holt, Rinehart and Winston. All rights reserved.

CONTENTS, CONTINUED

Copyright © by Holt, Rinehart and Winston. All rights reserved.

1.1 QUICK ACTIVITY, Section 1.2

Making Observations

(The activity corresponding to this datasheet begins on page 14 of the textbook.)

1. Get an ordinary candle of any shape and color.

2. Record all observations you can make about the candle.

_____

_____

_____

_____

_____

_____

_____

3. Light the candle and watch it burn for 1 minute.

4. Record as many observations about the burning candle as you can.

_____

_____

_____

_____

_____

_____

_____

_____

5. Share your results with your class, and find out how many different things were observed.

Copyright © by Holt, Rinehart and Winston. All rights reserved.

Making Measurements

(The lab corresponding to this datasheet begins on page 30 of the textbook.)

Table 1-6 Dimensions of a Rectangular Block

	Length (cm)	Width (cm)	Height (cm)	Volume (cm³)
Trial 1				
Trial 2				
Trial 3				
Average				

Table 1-7 Circumference of a Ball

	Circumference (cm)	Difference from average (cm)
Trial 1		
Trial 2		
Trial 3		
Average		

Table 1-8A Mass of Sodium Chloride

	Mass of beaker and sodium chloride (g)	Mass of beaker (g)	Mass of sodium chloride (g)
Trial 1			
Trial 2			
Trial 3			
Average			

Copyright © by Holt, Rinehart and Winston. All rights reserved.

Table 1-8B Mass of Sodium Hydrogen Carbonate

	Mass of beaker and sodium hydrogen carbonate (g)	Mass of beaker (g)	Mass of sodium hydrogen carbonate (g)
Trial 1			
Trial 2			
Trial 3			
Average			

Table 1-9 Liquid Volume

	Volume (mL)
Test tube 1	
Test tube 2	
Test tube 3	
Average	

Table 1-10 Volume of an Irregular Solid

	Total volume (mL)	Volume of water only (mL)	Volume of object (mL)
Trial 1			
Trial 2			
Trial 3			
Average			

Copyright © by Holt, Rinehart and Winston. All rights reserved.

Analyzing Your Results

1. On the grid below, make a line graph of the temperatures that were measured with the wall thermometer over time.

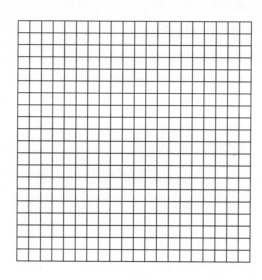

Time (hours)

Did the temperature change during the class period? If it did, find the average temperature, and determine the largest rise and the largest drop.

_____

_____

_____

Defending Your Conclusions

2. On the first grid below, make a bar graph using the data from the three calculations of the mass of sodium chloride. Indicate the average value of the three determinations by drawing a line that represents the average value across the individual bars. On the second grid below, do the same for the sodium hydrogen carbonate masses. Using the information in your graphs, determine whether you measured the sodium chloride or the sodium hydrogen carbonate more precisely.

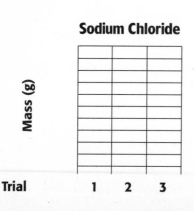

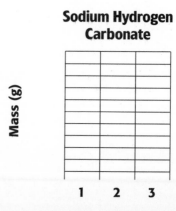

Copyright © by Holt, Rinehart and Winston. All rights reserved.

3. Suppose one of your test tubes has a capacity of 23 mL. You need to use about 5 mL of a liquid. Describe how you could estimate 5 mL.

_____

_____

_____

_____

4. Why is it better to align the meterstick with the edge of the object at the 1 cm mark rather than at the end of the stick?

_____

_____

_____

_____

5. Why is it better to place the meterstick on edge so that its scale is resting on the surface being measured rather than placing the meterstick flat?

_____

_____

_____

_____

6. Why do you think it is better to measure the circumference of the ball using string than to use a flexible metal measuring tape?

_____

_____

_____

_____

Copyright © by Holt, Rinehart and Winston. All rights reserved.

1.3 LABORATORY EXPERIMENT 1

Designing a Pendulum Clock

(The lab corresponding to this datasheet begins on page 1 of Laboratory Experiments.)

Pendulum Data

Independent variable (_____) Units	Period of the pendulum (s)			
	Trial 1	Trial 2	Trial 3	Average

Making a Hypothesis

Choose the one variable that you think will have the greatest effect on the period of the pendulum. Write a hypothesis that clearly states how changing this variable will affect the period of the pendulum. For example, if you increase your independent variable (length, mass, or distance), will it cause the period to be longer or shorter? Remember that this is only a *hypothesis*. If you wind up guessing wrong, that's okay. What matters is that you will test your hypothesis with an experiment to find out what is true.

_____

_____

_____

_____

Analyzing Your Results

1. Determine the average period of the pendulum for each set of trials by using the following equation. Be sure your answers have the proper number of significant figures. Record your answers in your data table.

$$\text{average period} = \frac{\text{trial 1 period} + \text{trial 2 period} + \text{trial 3 period}}{3}$$

Copyright © by Holt, Rinehart and Winston. All rights reserved.

2. Plot your data in the form of a graph on the grid below. Plot the independent variable on the *x*-axis, being sure to include what your independent variable was and what units it was measured in. Plot the average period (the dependent variable) on the *y*-axis. Connect the data points with the line or smooth curve that fits the points best. If you are using a graphing calculator, copy the graph from the calculator onto this datasheet.

**Relating Your Independent Variable to the
Period of the Pendulum**

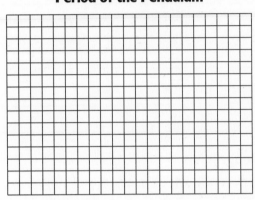

Average period (s)

Independent variable (units)

_____ (___)

Reaching Conclusions

3. Do your results support your hypothesis? Refer to your graph to explain why or why not.

_____

_____

_____

_____

4. Could the pendulum clock you made be adjusted to measure time accurately to the nearest one-thousandth (0.001) of a second? Explain why or why not.

_____

_____

_____

_____

_____

Copyright © by Holt, Rinehart and Winston. All rights reserved.

Defending Your Conclusions

5. Suppose someone tells you that if your results are precise, then they must be accurate. Is this a true statement? Explain why or why not.

_____

_____

_____

_____

_____

_____

Expanding Your Knowledge

1. Research several different devices used to measure time, and compare the accuracy and precision of the devices. Summarize your findings in the space below.

_____

_____

_____

_____

_____

_____

_____

_____

2. In the space below, outline your plan to design a pendulum clock that can accurately measure 1 minute. Compete with other members of your class to see who can develop the most accurate clock.

_____

_____

_____

_____

_____

_____

Copyright © by Holt, Rinehart and Winston. All rights reserved.

2.1 QUICK ACTIVITY, Section 2.2

Kinetic Theory

(The activity corresponding to this datasheet begins on page 48 of the textbook.)

You will need water, vegetable oil, and rubbing alcohol.

1. Dip one index finger into the water. Dip your other index finger into the oil. Wave each finger in the air. Do your fingers feel cool? How quickly does each liquid evaporate?

_____

_____

_____

_____

2. Repeat the experiment, using water on one finger and rubbing alcohol on the other.

3. Which of the three liquids evaporates the quickest? the slowest? Which liquid cools your skin the most? the least?

_____

_____

_____

_____

4. Use kinetic theory to explain your observations.

_____

_____

_____

_____

_____

_____

Copyright © by Holt, Rinehart and Winston. All rights reserved.

2.2 INQUIRY LAB, SECTION 2.3

How are the mass and volume of a substance related?

(The lab corresponding to this datasheet begins on page 57 of the textbook.)

Volume of H_2O (mL)	Mass of cylinder and H_2O (g)	Mass of H_2O (g)
0		—
10		
20		
30		
40		
50		
60		
70		
80		
90		
100		

Copyright © by Holt, Rinehart and Winston. All rights reserved.

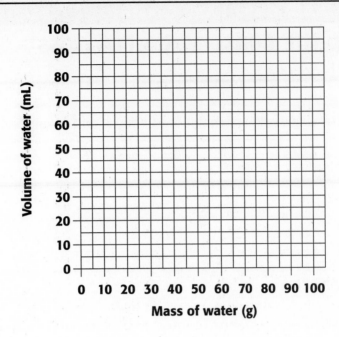

Analysis

1. Use your graph to predict the mass of 55 mL of water. What is the mass of 100 mL of water?

_____

_____

2. Use your graph to predict the volume of 25 g of water. What is the volume of 75 g of water?

_____

_____

3. How could you use your data table to calculate the density of water? How could you use your graph to calculate the density of water? Which method do you think gives better results? Why?

_____

_____

_____

_____

_____

_____

_____

Copyright © by Holt, Rinehart and Winston. All rights reserved.

Testing the Conservation of Mass

(The lab corresponding to this datasheet begins on page 64 of the textbook.)

Observing the Reaction Between Vinegar and Baking Soda

	Initial mass (g)	Final mass (g)	Change in mass (g)
Trial 1			
Trial 2			

Designing Your Experiment

8. Examine the plastic bag and the twist ties. With your lab partners, develop a procedure that will test the law of conservation of mass more accurately than trial 1 did. Which products' masses were not measured? How can you be sure you measure the masses of all of the reaction products?

9. In the space below, list each step you will perform in your experiment.

_____

_____

_____

_____

_____

_____

_____

_____

_____

_____

_____

_____

_____

10. Before you carry out your experiment, your teacher must approve your plan.

Copyright © by Holt, Rinehart and Winston. All rights reserved.

Analyzing Your Results

1. Compare the changes in mass you calculated for the first and second trials. What value would you expect to obtain for a change in mass if both trials validated the law of conservation of mass?

_____

_____

_____

_____

2. Was the law of conservation of mass violated in the first trial? Explain your reasoning.

_____

_____

_____

_____

3. If the results of the second trial were different from those of the first trial, explain why.

_____

_____

_____

_____

Defending Your Conclusions

4. Suppose someone performs an experiment like the one you designed and finds that the final mass is much less than the initial mass. Would that prove that the law of conservation of mass is wrong? Explain your reasoning.

_____

_____

_____

_____

Copyright © by Holt, Rinehart and Winston. All rights reserved.

Comparing the Buoyancy of Different Objects

(The lab corresponding to this datasheet begins on page 6 of Laboratory Experiments.)

Table 1 Jar Volumes

Jar	Volume of water (mL)	Volume of water and jar (mL)	Volume of jar (mL)
1			
2			

Table 2 Forces Acting on the Jars

Jar	Trial	Force acting on jar in air (N)	Force acting on jar underwater (N)	Buoyant force acting on jar (N)	Average buoyant force (N)
1	1				
	2				
	3				
2	1				
	2				
	3				

Copyright © by Holt, Rinehart and Winston. All rights reserved.

Copyright © by Holt, Rinehart and Winston. All rights reserved.

Analyzing Your Results

1. Calculate the volume of each jar with the following equation. Record your answers in **Table 1.**

$$\textbf{volume of jar} = \textbf{volume of water and jar} - \textbf{volume of water}$$

2. Calculate the buoyant force for each jar for each trial with the following equation. Record your answers in **Table 2.**

$$\textbf{buoyant force acting on jar} = \textbf{force acting in air} - \textbf{force acting in water}$$

3. Calculate the average buoyant force for each jar by adding the three buoyant forces and dividing by 3. Record your answers in **Table 2.**

4. Plot your data and the data from another group in the form of a graph on the grid below.

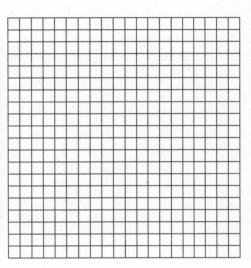

How Volume Affects Buoyant Force

Buoyant force (N)

Volume of jar (mL)

Reaching Conclusions

5. How is an object's buoyancy affected by its volume? How does an object's mass affect its buoyancy?

6. Use your graph to determine what size float the owner should buy to provide 5.0 N of buoyant force.

Defending Your Conclusions

7. Many floating objects are less dense than water. But boats are made of materials that are denser than water. What can you conclude about the buoyant force acting on a boat compared with the force of gravity acting on the boat?

Expanding Your Knowledge

1. Design your own boat using thick aluminum foil. Compete with your classmates to see whose boat can carry the most pennies before sinking. Outline the factors you will consider when designing your boat in the space below.

Copyright © by Holt, Rinehart and Winston. All rights reserved.

3.1 QUICK ACTIVITY, SECTION 3.1

Convincing John Dalton

(The activity corresponding to this datasheet begins on page 73 of the textbook.)

If Dalton were still alive, he might argue: "Atoms are neutral, so they can't be made of charged particles." Explain why this statement is not true.

_____

_____

_____

_____

_____

_____

Copyright © by Holt, Rinehart and Winston. All rights reserved.

CHAPTER 3

3.2 QUICK ACTIVITY, SECTION 3.1

Constructing a Model

(The activity corresponding to this datasheet begins on page 75 of the textbook.)

Characteristics	Observations
Mass	
Size	
Shape	
Texture	
Other observations	

Use the data you have collected to draw a model of the unknown object in the space below.

What is the actual object in the sock? How well does your model match the object?

_____

_____

_____

_____

Copyright © by Holt, Rinehart and Winston. All rights reserved.

Isotopes

(The activity corresponding to this datasheet begins on page 84 of the textbook.)

Calculate the number of neutrons there are in the following isotopes. Write your answers in the table below. (Use the periodic table to find the atomic numbers.)

Isotope	Atomic number	Number of neutrons
Carbon-14		
Nitrogen-15		
Sulfur-35		
Calcium-45		
Iodine-131		

Copyright © by Holt, Rinehart and Winston. All rights reserved.

Elements in Your Food

(The activity corresponding to this datasheet begins on page 88 of the textbook.)

1. For one day, make a list of the ingredients in all the foods and drinks you consume.

2. Identify which ingredients on your list are compounds.

3. For each compound on your list, try to figure out what elements it is made of.

Record your answers in the table below.

Ingredient	Compound? (yes or no)	Elements in compound

Copyright © by Holt, Rinehart and Winston. All rights reserved.

Why do some metals cost more than others?

(The lab corresponding to this datasheet begins on page 90 of the textbook.)

Metal	Abundance in Earth's crust (%)	Price ($/kg)
Aluminum (Al)	8.2	1.55
Chromium (Cr)	0.01	0.06
Copper (Cu)	0.0060	2.44
Gold (Au)	0.000 0004	11 666.53
Iron (Fe)	5.6	0.03
Silver (Ag)	0.000 007	154.97
Tin (Sn)	0.0002	6.22
Zinc (Zn)	0.007	1.29

1. The table above gives the abundance of some metals in Earth's crust. List the metals in order from the most abundant to the least abundant.

_____

_____

2. List the metals in order of price, from the cheapest to the most expensive.

_____

_____

3. If the price of a metal depends on its abundance, you would expect the order to be the same on both lists. How well do the two lists match? Mention any exceptions.

_____

_____

_____

4. The order of reactivity of these metals, from most reactive to least reactive, is aluminum, zinc, chromium, iron, tin, copper, silver, and gold. Use this information to explain any exceptions you noticed in item 3.

_____

_____

_____

Copyright © by Holt, Rinehart and Winston. All rights reserved.

CHAPTER 3

Comparing the Physical Properties of Elements

(The lab corresponding to this datasheet begins on page 104 of the textbook.)

Physical Properties of Some Metals

Metal	Density (g/mL)	Relative hardness	Relative heat conductivity	Magnetized by magnet?
Aluminum (Al)	2.7	28	100	No
Iron (Fe)	7.9	50	34	Yes
Nickel (Ni)	8.9	67	38	Yes
Tin (Sn)	7.3	19	28	No
Tungsten (W)	19.3	100	73	No
Zinc (Zn)	7.1	28	49	No

Designing Your Experiment

6. With your lab partner(s), decide how you will use the materials provided to identify each metal you are given. There is more than one way to measure some of the physical properties that are listed, so you might not use all of the materials that are provided.

7. In the space below, list each step you will perform in your experiment.

_____

_____

_____

_____

_____

_____

_____

_____

_____

_____

_____

_____

8. Have your teacher approve your plan before you carry out your experiment.

Copyright © by Holt, Rinehart and Winston. All rights reserved.

Analyzing Your Results

1. Fill in the table below by listing the physical properties you compared and the data you collected for each of the unknown metals.

Unknown metal	Properties tested		
	_____	_____	_____
1			
2			
3			

2. Which metals were you given? Explain the reasoning you used to identify each metal.

_____

_____

_____

_____

_____

_____

3. Which physical properties were the easiest for you to measure and compare? Which were the hardest? Explain why.

_____

_____

_____

_____

_____

_____

_____

Copyright © by Holt, Rinehart and Winston. All rights reserved.

4. What would happen if you tried to scratch aluminum foil with zinc?

_____

_____

5. Explain why it would be difficult to distinguish between iron and nickel unless you calculate each metal's density.

_____

_____

_____

_____

6. Suppose you find a metal fastener and determine that its density is 7 g/mL. What are two ways you could determine whether the unknown metal is tin or zinc?

_____

_____

_____

_____

Defending Your Conclusions

7. Suppose someone gives you an alloy that is made of both zinc and nickel. In general, how do you think the physical properties of the alloy would compare with those of each individual metal?

_____

_____

_____

_____

_____

Copyright © by Holt, Rinehart and Winston. All rights reserved.

3.7 **LABORATORY EXPERIMENT 3**

Predicting the Physical and Chemical Properties of Elements

(The lab corresponding to this datasheet begins on page 10 of Laboratory Experiments.*)*

Performing the Experiment

If you are not using a graphing calculator, you may plot the data on the grid below.

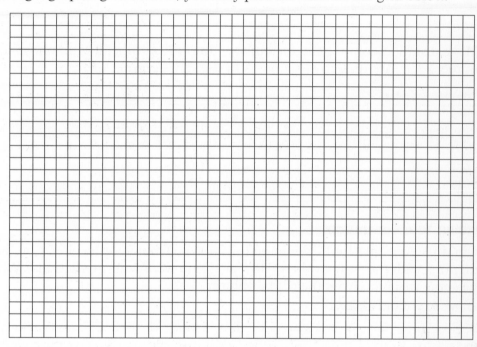

Property (___) (Units)

Atomic number

Analyzing Your Results

1. After studying the graph and reviewing the data in the table, predict the value of the property for the element with atomic number 3 (lithium). Enter your predicted value in the table below. To which family does lithium belong?

_____

2. Make the same prediction for the element with atomic number 9 (fluorine), and enter your prediction in the table below. To which family does fluorine belong?

_____

Some Properties of the Elements

Element	Atomic number	Density (g/mL)	Melting point (°C)	Boiling point (°C)	Charge of ion
Li	3				
F	9				

Reaching Conclusions

3. Are carbon (atomic number 6) and silicon (atomic number 14) in the same group of the periodic table? Compare the properties of these two elements.

_____

_____

_____

_____

4. Make a similar comparison for nitrogen (atomic number 7) and phosphorus (atomic number 15). Are these two elements in the same group?

_____

_____

_____

_____

Defending Your Conclusions

5. Obtain actual data for lithium and for fluorine from your teacher. Compare your predictions with the actual values. How accurate were your predictions? Why might there be some variation from the actual values?

_____

_____

_____

_____

_____

_____

Expanding Your Knowledge

1. Pick an element with an atomic number greater than 20, and research it. Find out when and how it was discovered, as well as who discovered it. Also find out its physical and chemical properties, its abundance on Earth, and any current uses it has in manufacturing or technology. Present your findings to the class.

Copyright © by Holt, Rinehart and Winston. All rights reserved.

4.1 INQUIRY LAB, SECTION 4.1

Which melts more easily, sugar or salt?

(The lab corresponding to this datasheet begins on page 113 of the textbook.)

Procedure

1. Use your knowledge of structures to make a hypothesis about whether sugar or salt will melt more easily.

_____

_____

_____

_____

_____

_____

Analysis

1. Which compound is easier to melt? Was your hypothesis right?

_____

_____

_____

_____

_____

2. How can you relate your results to the structure of each compound?

_____

_____

_____

_____

_____

Copyright © by Holt, Rinehart and Winston. All rights reserved.

4.2 **QUICK ACTIVITY, SECTION 4.2**

Building a Close-Packed Structure

(The activity corresponding to this datasheet begins on page 118 of the textbook.)

Copper and other metals have close-packed structures. This means their atoms are packed very tightly together. In this activity, you will build a close-packed structure using ping-pong balls.

1. Place three books flat on a table so that their edges form a triangle.

2. Fill the triangular space between the books with spherical "atoms." Adjust the books so that the atoms make a one-layer close-packed pattern.

3. Build additional layers on top of the first layer. How many other atoms does each atom touch? Where have you seen other arrangements that are similar to this one?

_____

_____

_____

_____

_____

_____

_____

_____

Copyright © by Holt, Rinehart and Winston. All rights reserved.

4.3 QUICK ACTIVITY, SECTION 4.4

Polymer Memory

(The activity corresponding to this datasheet begins on page 133 of the textbook.)

Polymers that return to their original shape after stretching can be thought of as having a "memory." In this activity, you will compare the memory of a rubber band with that of the plastic rings that hold a six-pack of cans together.

1. Which polymer stretches better without breaking?

_____

_____

2. Which one has better memory?

_____

_____

_____

3. Warm the stretched six-pack holder over a hot plate, being careful not to melt it. Does it retain its memory?

_____

_____

_____

_____

Copyright © by Holt, Rinehart and Winston. All rights reserved.

CHAPTER 4

4.4 INQUIRY LAB, Section 4.4
What properties does a polymer have?

(The lab corresponding to this datasheet begins on page 135 of the textbook.)

Analysis

1. Do both liquids take the shape of the pans? What does this mean?

_____

_____

_____

_____

2. What happened to the liquid made of sugar and water when it was hit with the mallet? What do you think happened to the molecules of this liquid when you hit the pan?

_____

_____

_____

_____

3. What happened to the liquid made of cornstarch and water when it was hit with the mallet? What happened to the mallet when it hit the liquid? What do you think happened to the molecules of this liquid when you hit it?

_____

_____

_____

_____

4. Which liquid has properties of a polymer? Explain how you reached this conclusion.

_____

_____

_____

_____

_____

Copyright © by Holt, Rinehart and Winston. All rights reserved.

4.5 SKILL BUILDER LAB, CHAPTER 4

Comparing Polymers

(The lab corresponding to this datasheet begins on page 140 of the textbook.)

Bounce Heights of Polymers

	Bounce height (cm)					
	Trial 1	**Trial 2**	**Trial 3**	**Trial 4**	**Trial 5**	**Average**
Latex rubber						
Ethanol-silicate						

Examine both polymers closely. Record how the two polymers are alike and how they are different.

_____

_____

_____

_____

Analyzing Your Results

1. Calculate the average bounce height for each ball by adding the five bounce heights and dividing by 5. Record the averages in your data table.

2. Based on only their bounce heights, which polymer would make a better toy ball?

_____

Defending Your Conclusions

3. Suppose that making a latex rubber ball costs 22 cents and that making an ethanol-silicate ball costs 25 cents. Does this fact affect your conclusion about which polymer would make a better toy ball? Besides cost, what are other important factors that should be considered?

_____

_____

_____

_____

Copyright © by Holt, Rinehart and Winston. All rights reserved.

Determining Which Household Solutions Conduct Electricity

(The lab corresponding to this datasheet begins on page 13 of Laboratory Experiments.)*

Table 1 The Conductivity of Some Household Solutions

Beaker number	Household solution	Conductivity (μS/cm)	Ionic, covalent, or polyatomic?
1	Tap water		
2	Table sugar		
3	Table salt		
4	Rubbing alcohol		
5	Milk of magnesia		
6	Epsom salt		
7	Baking soda		
8	Dishwashing soap		
9	Chemical fertilizer		

Analyzing Your Results

1. Complete the table below by putting each solution you tested into one of the three categories.

Table 2 Classifying Household Solutions

No conductivity (0 μS/cm)	
Low conductivity (1–500 μS/cm)	
High conductivity (> 500 μS/cm)	

Copyright © by Holt, Rinehart and Winston. All rights reserved.

Reaching Conclusions

2. Which household solutions might be contributing to the high conductivity of the river? Explain why.

_____

_____

_____

_____

_____

3. Explain why tap water shows some conductivity.

_____

_____

_____

_____

4. Which household solution has the highest conductivity?

_____

_____

5. Given the fact that there is a large golf course and several large gardens that have to be maintained in the vicinity of the river, which household item is probably the greatest contributor to the high conductivity of the river?

_____

_____

_____

Copyright © by Holt, Rinehart and Winston. All rights reserved.

Chemical Formulas of Some Household Substances

Household substance	Chemical formula and additional comments
Tap water	H_2O with some dissolved minerals
Table sugar	$C_{12}H_{22}O_{11}$ (organic compound)
Table salt	NaCl
Rubbing alcohol	C_3H_8O (organic compound) dissolved in water
Milk of magnesia	$Mg(OH)_2$ suspended in water
Epsom salt	$MgSO_4$
Baking soda	$NaHCO_3$
Dishwashing soap	Complex organic mixture; main ingredient is sodium lauryl sulfate, $C_{12}H_{25}SO_4Na$
Chemical fertilizer	Complex mixture of nutrients and minerals; includes nitrates, such as NO_3^-, and phosphates, such as PO_4^{3-}

6. The table above lists the chemical formulas and some comments for the substances in each solution you tested. Use your data and this information to make a general conclusion about the ability of solutions of organic compounds to conduct electricity. Are there any exceptions to your general conclusion? Explain.

_____

_____

_____

_____

_____

_____

7. For each solution you tested, use your data, your textbook, and the information provided in the table above to determine whether each substance has ionic or covalent bonds or contains one or more polyatomic ions. Record your answers in **Table 1.**

Copyright © by Holt, Rinehart and Winston. All rights reserved.

Defending Your Conclusions

8. Write a brief recommendation to the city council with some suggestions about what homeowners and local businesses can do to reduce the conductivity of the river. Explain to them why following your recommendations would help reduce the river's conductivity.

_____

_____

_____

_____

_____

Expanding Your Knowledge

1. Collect samples from several water sources in your area, such as wells, lakes, rivers, streams, and bayous, and determine each sample's conductivity. Contact local water-quality experts to find out what kinds of substances may contribute to the increased conductivity of the water in your area. Summarize your findings below.

_____

_____

_____

_____

_____

2. Develop a plan to test the conductivity of soil. Compare the conductivities of several samples of soil from different locations. Outline your plan below.

_____

_____

_____

_____

Copyright © by Holt, Rinehart and Winston. All rights reserved.

CHAPTER 4

5.1 QUICK ACTIVITY, SECTION 5.3

Candy Chemistry

(The activity corresponding to this datasheet begins on page 166 of the textbook.)

Look at the partial equations below. Using different-colored gumdrops to show atoms of different elements, make models of the reactions by connecting the "atoms" with toothpicks. Use your models to help you balance the following equations. Classify each reaction.

a. $C_3H_8 + O_2 \rightarrow CO_2 + H_2O$

Balanced equation:

_____ $C_3H_8 +$ _____ $O_2 \rightarrow$ _____ $CO_2 +$ _____ H_2O

Reaction type: _____

b. $KI + Br_2 \rightarrow KBr + I_2$

Balanced equation:

_____ $KI +$ _____ $Br_2 \rightarrow$ _____ $KBr +$ _____ I_2

Reaction type: _____

c. $H_2 + Cl_2 \rightarrow HCl$

Balanced equation:

_____ $H_2 +$ _____ $Cl_2 \rightarrow$ _____ HCl

Reaction type: _____

d. $FeS + HCl \rightarrow FeCl_2 + H_2S$

Balanced equation:

_____ $FeS +$ _____ $HCl \rightarrow$ _____ $FeCl_2 +$ _____ H_2S

Reaction type: _____

Copyright © by Holt, Rinehart and Winston. All rights reserved.

5.2 INQUIRY LAB, Section 5.3
Can you write balanced chemical equations?

(The lab corresponding to this datasheet begins on page 167 of the textbook.)

Analysis

1. What did you observe as a sign that a double-displacement reaction was occurring?

_____

_____

_____

2. Identify the reactants and products for each reaction.

_____

_____

_____

3. Write the balanced equation for each reaction.

_____

_____

_____

4. Which ion(s) produced a solid with silver nitrate?

_____

5. Does this test let you identify all the ions? Why or why not?

_____

_____

_____

Copyright © by Holt, Rinehart and Winston. All rights reserved.

5.3 INQUIRY LAB, SECTION 5.4

What affects the rates of chemical reactions?

(The lab corresponding to this datasheet begins on page 172 of the textbook.)

Analysis

1. Describe and interpret your results.

Magnesium, copper, and zinc in vinegar

_____

_____

_____

_____

Paper clip and steel wool in the hottest part of the flame

_____

_____

_____

_____

Magnesium in pure vinegar and diluted vinegar

_____

_____

_____

_____

Igniting a sugar cube alone and a cube with ash

_____

_____

_____

_____

Copyright © by Holt, Rinehart and Winston. All rights reserved.

2. For each step, list the factor(s) that influenced the rate of the reaction.

Magnesium, copper, and zinc in vinegar

_____

_____

_____

Paper clip and steel wool in the hottest part of the flame

_____

_____

_____

Magnesium in pure vinegar and diluted vinegar

_____

_____

_____

Igniting a sugar cube alone and a cube with ash

_____

_____

_____

Copyright © by Holt, Rinehart and Winston. All rights reserved.

5.4 QUICK ACTIVITY, Section 5.4

Catalysts in Action

(The activity corresponding to this datasheet begins on page 173 of the textbook.)

1. Pour 2 percent hydrogen peroxide into a test tube to a depth of 2 cm.

2. Pour the same volume of water into another test tube.

3. Drop a small piece of raw liver into each test tube.

4. Liver contains the enzyme catalase. Watch carefully, and describe what happens. Explain your observations.

_____

_____

_____

_____

_____

5. Repeat steps 1–4 using a piece of liver that has been boiled for 3 minutes. Explain your result.

_____

_____

_____

_____

6. Repeat steps 1–4 again using iron filings instead of liver. What happens?

_____

_____

_____

_____

Copyright © by Holt, Rinehart and Winston. All rights reserved.

Measuring the Rate of a Chemical Reaction

(The lab corresponding to this datasheet begins on page 180 of the textbook.)

	Mass of metal (g)	Diameter or width of metal (mm)	Length of metal (mm)	Surface area of metal (mm²)	Reaction time (s)
Reaction 1					
Reaction 2					

Designing Your Experiment

7. By completing steps 1–6, you have half the data you need to answer the question posed at the beginning of the lab. How can you use the aluminum foil to collect the rest of the data? You will need to adapt the procedure to measure how long the reaction takes and to determine the surface area of the aluminum foil.

8. In the space below, list each step you will perform in your experiment. Because the surface area is the variable you want to test, the other factors should be the same as in steps 1–6.

_____

_____

_____

_____

_____

_____

_____

_____

_____

_____

_____

_____

_____

_____

9. Before you carry out your experiment, your teacher must approve your plan.

Copyright © by Holt, Rinehart and Winston. All rights reserved.

Analyzing Your Results

1. Calculate the surface area of the piece of wire by using the equation below. Record the result in your data table.

$$\text{area (in mm}^2) = \left(\frac{\pi \times \text{diameter}^2}{2}\right) + (\pi \times \text{diameter} \times \text{length})$$

2. Calculate the surface area of the foil by using the equation below. Record the result in your data table.

$$\text{foil area (in mm}^2) = 2 \times \text{width of foil} \times \text{length of foil}$$

3. Which has the larger surface area, the wire or the foil? Calculate the ratio of the surface areas by dividing the larger surface area by the smaller one.

4. Which reacted more rapidly with hydrochloric acid, the wire or the foil?

5. How does increasing the surface area of a reactant affect the rate of this reaction?

Defending Your Conclusions

6. If someone tells you that the reaction rates differed because the wire or the foil was not pure aluminum, how could you do this experiment using either wire or foil but not both?

Copyright © by Holt, Rinehart and Winston. All rights reserved.

5.6 LABORATORY EXPERIMENT 5

Investigating the Effect of Temperature on the Rate of a Reaction

(The lab corresponding to this datasheet begins on page 17 of Laboratory Experiments.*)*

Temperature and Reaction Time of Reactants

Trial	Temperature (°C)	Elapsed reaction time (s)
1		
2		
3		
4		

Analyzing Your Results

1. Plot your data on the graph below. If you are using a graphing calculator, copy the graph from the calculator onto this datasheet.

Relating Reaction Time to Temperature

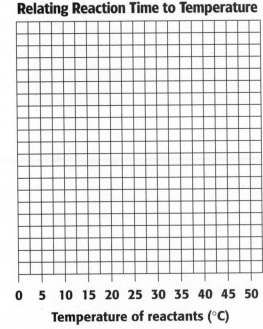

Elapsed reaction time (s)

0 5 10 15 20 25 30 35 40 45 50

Temperature of reactants (°C)

2. Describe how the temperature of the reactants affects the time it takes for this reaction to occur.

_____

_____

_____

Copyright © by Holt, Rinehart and Winston. All rights reserved.

3. Use the graph to estimate the time it would take for this reaction to occur if the reactants were at 5°C. What would the reaction time be if the reactants were at 50°C?

_____

_____

4. Do you think the relationship shown by the graph applies to temperatures below 0°C? (**Hint:** What would happen to the water in the solutions below 0°C?)

_____

_____

_____

_____

Reaching Conclusions

5. Assuming that the reaction of sodium thiosulfate with silver bromide is similar to the reaction you carried out in this experiment, what is the minimum temperature your chemicals must be to ensure that the reaction is complete in 20 s or less?

_____

_____

6. Considering that some of your photo processing chemicals may undergo decomposition reactions when they are stored at room temperature, what is one thing you can do to slow these reactions and extend the shelf life of your chemicals?

_____

_____

_____

Copyright © by Holt, Rinehart and Winston. All rights reserved.

Defending Your Conclusions

7. Suppose someone tells you your results are not valid because you did not test the actual reaction used to develop film. What can you do to show your results are valid?

_____

_____

_____

_____

_____

Expanding Your Knowledge

1. Another factor that affects the rate of a reaction is how dilute a solution of a reactant is. To test the effects of this factor, repeat the same reaction at room temperature, varying the concentration of HCl each time. For the first trial, combine 5 mL of each solution. For the next trial, add 3 mL of deionized water to 2 mL of the HCl solution, and combine this solution with 5 mL of $Na_2S_2O_3$. For the last trial, add 4 mL of deionized water to 1 mL of the HCl solution, and combine this solution with 5 mL of $Na_2S_2O_3$. Make a graph of your data on the grid below.

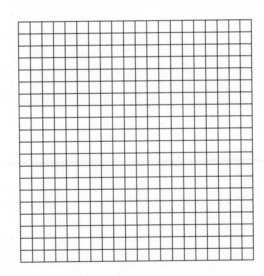

Elapsed reaction time (s)

Volume of original HCl solution added (mL)

Explain how changing this factor affects the rate of the reaction.

_____

_____

_____

Copyright © by Holt, Rinehart and Winston. All rights reserved.

6.1 QUICK ACTIVITY, SECTION 6.1

Modeling Decay and Half-life

(The activity corresponding to this datasheet begins on page 192 of the textbook.)

Trial	Number of heads	Number of tails	Ratio of heads/number used
1			
2			
3			
4			
5			
6			
7			
8			
9			
10			
11			
12			
13			
14			
15			

4. For each trial, divide the number of heads-up pennies set aside by the total number of pennies used in the trial. Are these ratios nearly equal to each other? What fraction are they closest to?

_____

_____

_____

_____

Copyright © by Holt, Rinehart and Winston. All rights reserved.

6.2 QUICK ACTIVITY, Section 6.2

Modeling Chain Reactions

(The activity corresponding to this datasheet begins on page 199 of the textbook.)

1. To model a fission chain reaction, you will need a small wooden building block and a set of dominoes.

2. Place the building block on a table or counter. Stand one domino upright in front of the block and parallel to one of its sides, as shown on page 199 of the textbook.

3. Stand two more dominoes vertically, parallel, and symmetrical to the first domino. Continue this process until you have used all the dominoes and a triangular shape is created, as shown on page 199 of the textbook.

4. Gently push the first domino away from the block so that it will fall over and hit the second group. How many dominoes fall over with each step?

_____

_____

_____

Copyright © by Holt, Rinehart and Winston. All rights reserved.

6.3 SKILL BUILDER LAB, CHAPTER 6
Simulating Nuclear Decay Reactions

(The lab corresponding to this datasheet begins on page 210 of the textbook.)

Throw number	Number of dice representing each isotope			
	$^{210}_{82}$Pb	$^{210}_{83}$Bi	$^{210}_{84}$Po	$^{206}_{82}$Pb
0 (start)	10	0	0	0
1				
2				
3				
4				
5				
6				
7				
8				
9				
10				
11				
12				

Analyzing Your Results

1. Write nuclear decay equations for the nuclear reactions modeled in this lab.

_____

_____

_____

Copyright © by Holt, Rinehart and Winston. All rights reserved.

2. Using a different color or symbol for each isotope, plot the data for all four isotopes on the graph below.

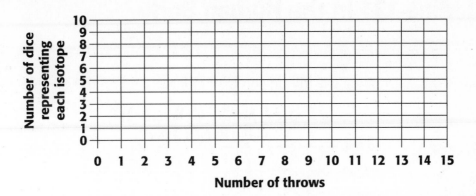

3. What do your results suggest about how the amounts of $^{210}_{82}Pb$ and $^{206}_{82}Pb$ on Earth are changing over time?

_____

_____

_____

_____

Defending Your Conclusions

4. $^{210}_{82}Pb$ is continually produced through a series of nuclear decays that begin with $^{238}_{92}U$. Does this information cause you to modify your answer to item 3? Explain why.

_____

_____

_____

_____

_____

_____

Copyright © by Holt, Rinehart and Winston. All rights reserved.

6.4 LABORATORY EXPERIMENT 6

Determining the Effective Half-life of Iodine-131 in the Human Body

(The lab corresponding to this datasheet begins on page 21 of Laboratory Experiments.*)*

If you are not using a graphing calculator, you may plot the data on the graph below.

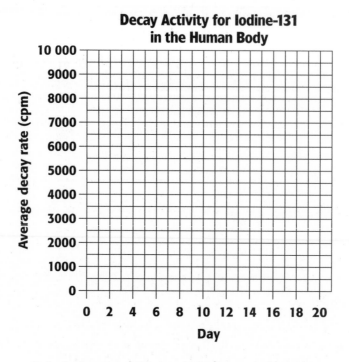

Calculating the Effective Half-life of Iodine-131

Decay rate (cpm)	Time (days)	Elapsed time to halve the decay rate (days)
10 000		
5000		
2500		
1250		
Average effective half-life		

Analyzing Your Results

1. Use your graph to describe how the concentration of iodine in the thyroid glands of the patients changed over the 20-day research period.

_____

_____

Copyright © by Holt, Rinehart and Winston. All rights reserved.

2. When did the thyroid glands of the patients have the maximum concentration of iodine-131?

_____

_____

_____

_____

3. Use your graph to estimate (to the nearest 0.1 day) the time at which the average decay rate was 10 000 cpm. Enter your answer in your data table. Do the same to estimate the time at which the average decay rate was 5000 cpm, 2500 cpm, and 1250 cpm. Record your answers in your data table.

4. Calculate how long it took (to the nearest 0.1 day) for the average decay rate to drop from 10 000 cpm to 5000 cpm. Do this by subtracting the time at which the decay rate was 10 000 cpm from the time at which the decay rate was 5000 cpm. Record this value in your data table.

5. Repeat item 4 to find out how many days it took for the average decay rate to drop from 5000 cpm to 2500 cpm and then from 2500 cpm to 1250 cpm. Record each of your answers in your data table. These values and the value from item 4 represent effective half-lives for iodine-131 in the human body.

6. Calculate the average effective half-life of iodine-131 by adding the three effective half-lives you just found, and then dividing by 3. Record your answer in your data table.

Reaching Conclusions

7. Explain why the concentration of iodine-131 was low initially, increased quickly, and then slowly decreased over time.

_____

_____

_____

_____

8. The actual half-life of iodine-131 is 8.07 days. How does this value compare with the average effective half-life that you calculated? Why is the effective half-life different from the actual half-life?

_____

_____

_____

_____

Copyright © by Holt, Rinehart and Winston. All rights reserved.

9. If the liver absorbs one-tenth as much iodine as the thyroid gland, how would the graph have looked if the study had been done on the liver instead of the thyroid gland?

_____

_____

_____

_____

Defending Your Conclusions

10. In this study, the individuals tested each had similar results. But suppose the data had shown that the amount of iodine-131 in the thyroid varied considerably from person to person. Would these results change the way doctors used the data to determine a specific dose? Explain.

_____

_____

_____

_____

Expanding Your Knowledge

1. Research the effects of nuclear weapons testing and the accident at Chernobyl that occurred in the Ukraine in 1986. Make a poster outlining how iodine-131 makes its way through the environment, eventually reaching humans.

2. Research how spent fuel rods from nuclear power plants are disposed of. In groups of four, develop a proposal for the safe storage of waste generated from nuclear power plants. Prepare a panel discussion to present and defend your proposal to your classmates. Outline your proposal below.

_____

_____

_____

_____

_____

_____

Copyright © by Holt, Rinehart and Winston. All rights reserved.

7.1 QUICK ACTIVITY, Section 7.3

Newton's First Law

(The activity corresponding to this datasheet begins on page 235 of the textbook.)

1. Place an index card over a glass, and set a coin on top of the index card.

2. With your thumb and forefinger, quickly flick the card sideways off the glass. Observe what happens to the coin. Does the coin move with the index card?

_____

_____

_____

_____

3. Try again, but this time slowly pull the card sideways and observe what happens to the coin.

4. Use Newton's first law to explain your results.

_____

_____

_____

_____

_____

_____

Copyright © by Holt, Rinehart and Winston. All rights reserved.

CHAPTER 7

7.2 INQUIRY LAB, Section 7.3

How are action and reaction forces related?

(The lab corresponding to this datasheet begins on page 239 of the textbook.)

Analysis

1. What are the action and reaction forces involved in the spring scale-mass system you have constructed?

_____

_____

_____

_____

_____

2. How did the readings on the two spring scales in step 4 compare? Explain how this is an example of Newton's third law of motion.

_____

_____

_____

_____

_____

_____

Copyright © by Holt, Rinehart and Winston. All rights reserved.

7.3 DESIGN YOUR OWN LAB, CHAPTER 7

Measuring Forces

(The lab corresponding to this datasheet begins on page 244 of the textbook.)

Table 1 Calibration

Rubber-band length (cm)	Change in length (cm)	Mass on hook (g)	Mass on hook (kg)	Force (N)
	0	0	0	0

Designing Your Experiment

10. With your lab partner(s), devise a plan to measure the force required to break a human hair using the instrument you just calibrated. How will you attach the hair to your instrument? How will you apply force to the hair?

11. In the space below, list each step you will perform in your experiment.

_____

_____

_____

_____

_____

_____

_____

_____

_____

_____

12. Have your teacher approve your plan before you carry out your experiment.

Copyright © by Holt, Rinehart and Winston. All rights reserved.

Table 2 Experimentation

Trial	Rubber-band length (cm)	Force (N)
Hair 1		
Hair 2		
Hair 3		

Analyzing Your Results

1. Plot your calibration data in the form of a graph on the grid below. Connect your data points with the line or smooth curve that fits the points best.

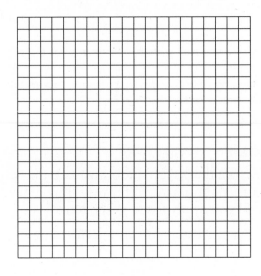

Force (N)

Length (cm)

2. Use the graph and the length of the rubber band for each trial of your experiment to determine the force that was necessary to break each of the three hairs. Record your answers in **Table 2.**

Defending Your Conclusions

3. Suppose someone tells you that your results are flawed because you measured length and not force. How can you show that your results are valid?

_____

_____

_____

_____

_____

_____

Copyright © by Holt, Rinehart and Winston. All rights reserved.

7.4 LABORATORY EXPERIMENT 7
Determining Your Acceleration on a Bicycle

(The lab corresponding to this datasheet begins on page 25 of Laboratory Experiments.)

Table 1 Forces and Masses of Objects

Object	Weight (lb)	Force (N)	Mass (kg)	
Bicycle and you				
You				Ratio of your force to hanging object's force
Hanging object				

Table 2 Force and Acceleration Measurements and Calculations

Front gear-wheel	Rear gear-wheel	Impelling force hanging object produces (N)	Your impelling force (N)	Your acceleration on the bicycle (m/s^2)
Smallest	S1 (low)			
	S2			
	S3			
	S4			
	S5			
Middle	M1 (low)			
	M2			
	M3			
	M4			
	M5			
Largest	L1 (low)			
	L2			
	L3			
	L4			
	L5			

Note: The number of gears on your bicycle might be different than the number shown in **Table 2.**

Copyright © by Holt, Rinehart and Winston. All rights reserved.

Analyzing Your Results

1. Convert both your weight and the combined weight of the bicycle and you to newtons by using the following equation. Record your answers in **Table 1.**

$$\text{force (N)} = \text{weight (lb)} \times \frac{1\text{ N}}{0.225\text{ lb}}$$

2. Calculate the combined mass of the bicycle and you in kilograms by using the following equation. Record your answer in **Table 1.**

$$\text{mass} = \frac{\text{force}}{\text{free-fall acceleration (9.8 m/s}^2)}$$

3. Calculate the downward force of the hanging object in newtons by using Newton's second law of motion. Record your answer in **Table 1.**

$$\text{force} = \text{mass} \times \text{free-fall acceleration (9.8 m/s}^2)$$

4. Assume that when you ride a bicycle you can use your entire weight to pedal. Calculate the ratio of your downward force to that of the hanging object by using the following equation. Record your answer in **Table 1.**

$$\text{ratio of your force to hanging object's force} = \frac{\text{your force}}{\text{force of hanging object}}$$

5. For each gear, calculate your impelling force by using the following equation. Record your answers in **Table 2.**

$$\text{your impelling force} = \text{object's impelling force} \times \text{ratio of forces}$$

6. For each gear, calculate your acceleration on the bicycle by using the following equation. Record your answers in **Table 2.**

$$\text{acceleration} = \frac{\text{your impelling force}}{\text{combined mass of bicycle and you}}$$

Copyright © by Holt, Rinehart and Winston. All rights reserved.

7. Plot your acceleration data from **Table 2** in the form of a bar graph on the grid below.

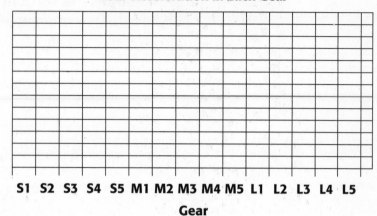

Your Acceleration in Each Gear

Acceleration (m/s²) vs. Gear (S1 S2 S3 S4 S5 M1 M2 M3 M4 M5 L1 L2 L3 L4 L5)

Reaching Conclusions

8. Use your bar graph to list the gears in order from the one that provides the most acceleration to the one that provides the least acceleration.

_____

_____

_____

9. Which gear(s) on your smallest front gear-wheel overlap with those on the middle front gear-wheel? Which gear(s) on the middle front gear-wheel overlap with those on the largest front gear-wheel?

_____

_____

_____

Defending Your Conclusions

10. In this experiment, you calculated the force you would exert on the bicycle by assuming that you could put all of your weight on the pedals. Is this a realistic assumption? How might your results have been different if you were able to measure the actual force you could exert while riding the bicycle?

_____

_____

_____

_____

Copyright © by Holt, Rinehart and Winston. All rights reserved.

Expanding Your Knowledge

1. One of the simplest ways bicyclists compare gears is by using the following equation.

$$\text{distance} = \text{diameter of rear wheel} \times \pi \times \frac{\text{number of teeth in front gear-wheel}}{\text{number of teeth in rear gear-wheel}}$$

This equation calculates the distance you can travel for each revolution of the pedals. Use this equation to calculate the distance in meters you could travel in each gear for a single turn of the pedals.

Gear	Distance traveled per turn (m)
S1 (low)	
S2	
S3	
S4	
S5	
M1 (low)	
M2	
M3	
M4	
M5	
L1 (low)	
L2	
L3	
L4	
L5	

Copyright © by Holt, Rinehart and Winston. All rights reserved.

Copyright © by Holt, Rinehart and Winston. All rights reserved.

8.1 INQUIRY LAB, SECTION 8.1

What is your power output when you climb the stairs?

(The activity corresponding to this datasheet begins on page 253 of the textbook.)

Weight or mass (lb or kg)	Weight (N)	Time to go up stairs (s)	Stair height (m)	Number of stairs	Total stair height (m)	Work (J)	Power (W)
You							
Your partner							

Procedure

1. Determine your weight in newtons. If your school has a scale that measures in kilograms, multiply your mass in kilograms by 9.8 m/s² to determine your weight in newtons. If your school has a scale that weighs in pounds, you can use the conversion factor of 4.45 N/lb.

2. Divide into pairs. Have your partner use the stopwatch to time how long it takes you to walk quickly up the stairs. Record the time in your data table. Then switch roles and repeat.

3. Measure the height of one step in meters. Multiply the number of steps by the height of one step to get the total height of the stairway. Record your answers in your data table.

4. Multiply your weight in newtons by the height of the stairs in meters to get the work you did in joules. Recall the work equation:

$$\text{work} = \text{force} \times \text{distance}$$

5. To get your power in watts, divide the work done in joules by the time in seconds that it took you to climb the stairs.

CHAPTER 8

Analysis

1. How would your power output change if you walked up the stairs faster?

_____

_____

2. What would your power output be if you climbed the same stairs in the same amount of time while carrying a stack of books weighing 20 N?

_____

_____

_____

3. Why did you use your weight as the force in the work equation?

_____

_____

_____

Copyright © by Holt, Rinehart and Winston. All rights reserved.

8.2 QUICK ACTIVITY, Section 8.2

A Simple Lever

(The activity corresponding to this datasheet begins on page 258 of the textbook.)

1. Make a first-class lever by placing a rigid ruler across a pencil or by crossing two pencils at right angles. Use this lever to lift a small stack of books.

2. Vary the location of the fulcrum and see how that affects the lifting strength. Why are the books easier to lift in some cases than in others?

_____

_____

_____

_____

_____

_____

_____

_____

Copyright © by Holt, Rinehart and Winston. All rights reserved.

8.3   QUICK ACTIVITY, SECTION 8.2

A Simple Inclined Plane

(The activity corresponding to this datasheet begins on page 260 of the textbook.)

1. Make an inclined plane out of a board and a stack of books.

2. Tie a string to an object that is heavy but has low friction, such as a metal toy car or a roll of wire. Use the string to pull the object up the plane.

3. Still using the string, try to lift the object straight up through the same distance.

4. Which action required more force? In which case did you do more work?

_____

_____

_____

_____

_____

_____

_____

Copyright © by Holt, Rinehart and Winston. All rights reserved.

8.4 QUICK ACTIVITY, Section 8.4

Energy Transfer

(The activity corresponding to this datasheet begins on page 275 of the textbook.)

1. Flex a piece of thick wire or part of a coat hanger back and forth about 10 times with your hands. Are you doing work?

_____

_____

_____

_____

2. After flexing the wire, cautiously touch the part of the wire where you bent it. Does the wire feel hot? What happened to the energy you put into it?

_____

_____

_____

_____

_____

_____

DATASHEET

8.5 INQUIRY LAB, SECTION 8.4

Is energy conserved in a pendulum?

(The lab corresponding to this datasheet begins on page 277 of the textbook.)

Procedure

1. Hang the pendulum bob from the string in front of a chalkboard. On the board, draw the diagram as shown in the textbook on page 277. Use the meterstick and the level to make sure the horizontal line is parallel to the ground.

2. Pull the pendulum back to the "X." Make sure everyone is out of the way; then release the pendulum and observe its motion. How high does the pendulum swing on the other side?

_____

_____

3. Let the pendulum swing back and forth several times. How many swings does the pendulum make before the ball noticeably fails to reach its original height?

_____

_____

4. Stop the pendulum and hold it again at the "X" marked on the board. Have another student place the eraser end of a pencil on the intersection of the horizontal and vertical lines. Make sure everyone is out of the way again, especially the student holding the pencil.

5. Release the pendulum again. This time its motion will be altered halfway through the swing as the string hits the pencil. How high does the pendulum swing now? Why?

_____

_____

_____

_____

_____

Copyright © by Holt, Rinehart and Winston. All rights reserved.

6. Try placing the pencil at different heights along the vertical line. How does this affect the motion of the pendulum? If you put the pencil down close enough to the arc of the pendulum, the pendulum will do a loop around it. Why does that happen?

_____

_____

_____

_____

_____

_____

Analysis

1. Use the law of conservation of energy to explain your observations in steps 2–6.

_____

_____

_____

_____

_____

_____

2. If you let the pendulum swing long enough, it will start to slow down, and it won't rise to the line any more. That suggests that the system has lost energy. Has it? Where did the energy go?

_____

_____

_____

_____

_____

_____

Copyright © by Holt, Rinehart and Winston. All rights reserved.

CHAPTER 8

8.6 SKILL BUILDER LAB, CHAPTER 8

Determining the Energy of a Rolling Ball

(The lab corresponding to this datasheet begins on page 284 of the textbook.)

	Height 1	Height 2	Height 3
Mass of ball (kg)			
Length of ramp (m)			
Height of ramp (m)			
Time ball traveled, first trial (s)			
Time ball traveled, second trial (s)			
Time ball traveled, third trial (s)			
Average time ball traveled (s)			
Final speed of ball (m/s)			
Final kinetic energy of ball (J)			
Initial potential energy of ball (J)			

Analyzing Your Results

1. Calculate the average speed of the ball using the following equation.

$$\text{average speed} = \frac{\text{length of ramp}}{\text{average time ball traveled}}$$

2. Multiply the average speed by 2 to obtain the final speed of the ball, and record the final speed.

3. Calculate and record the final kinetic energy of the ball by using the following equation.

$$KE = \frac{1}{2} \times \text{mass of ball} \times (\text{final speed})^2$$

Copyright © by Holt, Rinehart and Winston. All rights reserved.

4. Calculate and record the initial potential energy of the ball by using the following equation.

$$\text{grav. PE} = \text{mass of ball} \times 9.8 \text{ m/s}^2 \times \text{height of ramp}$$

Defending Your Conclusions

5. For each of the three heights, compare the ball's potential energy at the top of the ramp with its kinetic energy at the bottom of the ramp.

_____

_____

_____

_____

_____

_____

6. How did the ball's potential and kinetic energy change as the height of the ramp was increased?

_____

_____

_____

_____

7. Suppose you perform this experiment and find that your kinetic energy values are always just a little less than your potential energy values. Does that mean you did the experiment wrong? Why or why not?

_____

_____

_____

_____

_____

_____

Copyright © by Holt, Rinehart and Winston. All rights reserved.

CHAPTER 8

Determining Which Ramp Is More Efficient

(The lab corresponding to this datasheet begins on page 30 of Laboratory Experiments.)

Table 1 Work Output for Both Ramps

Height of chair (to seat) (m)	Weight of cart (N)	Work output (J)

Table 2 Work Input, Mechanical Advantage, and Efficiency of Each Ramp

Ramp length (m)	Sliding or rolling the cart?	Force (N)	Work input (J)	Mechanical advantage	Efficiency (%)
	Sliding				
	Rolling				
	Sliding				
	Rolling				

Analyzing Your Results

1. Calculate the work output for both ramps in joules by using the following equation. Record your answer in **Table 1.**

   work output = height of chair (to seat) × weight of cart

   This is the work required to lift the cart from the floor to the seat of the chair without using any machines and without any friction.

Copyright © by Holt, Rinehart and Winston. All rights reserved.

Copyright © by Holt, Rinehart and Winston. All rights reserved.

2. Use the following equation to calculate the work input required using each ramp, first when sliding and then when rolling the cart. Record your answers in **Table 2.**

$$\textbf{work input} = \textbf{force} \times \textbf{ramp length}$$

This is the work required to move the cart from the floor to the seat of the chair using the ramp.

3. Calculate the mechanical advantage of each ramp by using the following equation. Record your answers in **Table 2.**

$$\textbf{mechanical advantage} = \frac{\textbf{ramp length}}{\textbf{height of chair (to seat)}}$$

4. Use the following equation to calculate the percent efficiency of each ramp, first when sliding and then when rolling the cart. Record your answers in **Table 2.**

$$\textbf{percent efficiency} = \frac{\textbf{work output}}{\textbf{work input}} \times 100$$

Reaching Conclusions

5. Which ramp has a greater mechanical advantage? How does the length of a ramp affect its mechanical advantage? For the Speedy Shipping Company, would you recommend a ramp with a greater or a lesser mechanical advantage?

CHAPTER 8

6. Which ramp is more efficient? In your own words, explain to the supervisor why the ramp is more efficient.

_____

_____

_____

_____

7. In each case, the work output is less than the work input. Explain why.

_____

_____

_____

8. Suppose that a driver needs to load an appliance weighing 1010 N into the back of a truck 1.0 m off the ground. Speedy Shipping Company does not allow its employees to exert more than 250 N of force. How long of a ramp should the driver use? Assume that the driver can use a rolling cart to reduce friction.

_____

_____

Copyright © by Holt, Rinehart and Winston. All rights reserved.

Copyright © by Holt, Rinehart and Winston. All rights reserved.

Defending Your Conclusions

9. Suppose someone tells you that your results are not valid because you measured the force needed to pull a light rolling cart up the ramp instead of a heavy appliance. Are your results still valid? Explain your answer.

_____

_____

_____

_____

Expanding Your Knowledge

1. Research the efficiencies of several brands of refrigerators or other household appliances. Summarize your results below.

_____

_____

_____

_____

_____

_____

_____

_____

_____

_____

_____

_____

CHAPTER 8

9.1 QUICK ACTIVITY, SECTION 9.1

Sensing Hot and Cold

(The activity corresponding to this datasheet begins on page 291 of the textbook.)

For this exercise, you will need three bowls.

1. Put an equal amount of water in all three bowls. In the first bowl, put some cold tap water. Put some hot tap water in the second bowl. Then mix equal amounts of hot and cold tap water in the third bowl.

2. Place one hand in the hot water and the other hand in the cold water. Leave them there for 15 s.

3. Place both hands in the third bowl that contains the mixture of hot and cold water. How does the water temperature feel to each hand? Explain.

_____

_____

_____

_____

_____

_____

_____

_____

Copyright © by Holt, Rinehart and Winston. All rights reserved.

9.2 INQUIRY LAB, SECTION 9.1

How do temperature and energy relate?

(The lab corresponding to this datasheet begins on page 295 of the textbook.)

Procedure

	Initial temperature (°C)	Highest temperature reached (°C)
Cup 1		
Cup 2		

Analysis

1. Which cup had the higher final temperature?

_____

_____

2. Both cups had the same starting temperature. Both sets of washers started at 100°C. Why did one cup reach a higher final temperature?

_____

_____

_____

_____

_____

_____

_____

Copyright © by Holt, Rinehart and Winston. All rights reserved.

Convection

(The activity corresponding to this datasheet begins on page 299 of the textbook.)

Describe the motion of the tiny soot particles in the space below.

_____

_____

_____

_____

_____

_____

_____

_____

_____

Copyright © by Holt, Rinehart and Winston. All rights reserved.

What color absorbs more radiation?

(The lab corresponding to this datasheet begins on page 300 of the textbook.)

Time (min)	Temperature of painted can (°C)	Temperature of unpainted can (°C)
3		
6		
9		
12		
15		
18		

Analysis

1. Plot the data for each can on the grid below. Be sure to label both sets of data clearly.

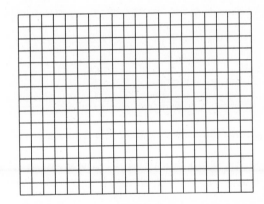

Temperature of can (°C)

Time (min)

2. Which color absorbed more radiation—silver or black?

_____

3. Which of the following variables in the lab were controlled (unchanged throughout the experiment)? Explain why each variable was or was not controlled.

a. starting temperature of water in cans

_____

_____

_____

Copyright © by Holt, Rinehart and Winston. All rights reserved.

b. volume of water in cans

_____

_____

_____

c. distance of cans from light

_____

_____

_____

d. size of cans

_____

_____

_____

4. Use your results to explain why panels used for solar heating are often painted black.

_____

_____

_____

_____

5. Based on your results, what color would you want your car to be in the winter? What color would you want your car to be in the summer? Justify your answers.

_____

_____

_____

_____

Copyright © by Holt, Rinehart and Winston. All rights reserved.

9.5 **QUICK ACTIVITY, SECTION 9.2**

Conductors and Insulators

(The activity corresponding to this datasheet begins on page 301 of the textbook.)

For this activity, you will need several flatware utensils. Each one should be made of a different material, such as stainless steel, aluminum, and plastic. You will also need a bowl and ice cubes.

1. Place the ice cubes in the bowl. Position the utensils in the bowl so that an equal length of each utensil lies under the ice.

2. Check the utensils' temperature every 20 s by briefly touching each utensil at the same distance from the ice that you touch each of the other utensils. Which utensil becomes the coldest first? What variables might affect your results?

_____

_____

_____

_____

_____

_____

_____

_____

_____

Copyright © by Holt, Rinehart and Winston. All rights reserved.

CHAPTER 9

9.6 DESIGN YOUR OWN LAB, CHAPTER 9

Investigating the Conduction of Heat

(The lab corresponding to this datasheet begins on page 316 of the textbook.)

Designing Your Experiment

5. With your lab partner(s), decide how you will use the materials available in the lab to compare the speed of conduction in three wires of different thicknesses. Do you think a thick wire will conduct energy more quickly or more slowly than a thin wire? Write your hypothesis in the space below.

_____

_____

_____

6. In the space below, list each step you will perform in your experiment.

_____

_____

_____

_____

_____

_____

_____

_____

_____

7. Before you carry out your experiment, your teacher must approve your plan.

	Wire diameter (mm)	Time to melt wax (s)			
		Trial 1	Trial 2	Trial 3	Average time
Wire 1					
Wire 2					
Wire 3					

Copyright © by Holt, Rinehart and Winston. All rights reserved.

Analyzing Your Results

1. Find the diameter of each wire you tested, and record each diameter in your data table in millimeters. If the diameter is listed in inches, convert it to millimeters by multiplying by 25.4. If the diameter is listed in mils, convert it to millimeters by multiplying by 0.0254.

2. Calculate the average time required to melt the ball of wax for each wire. Record your answers in your data table.

3. Plot your data in the form of a graph on the grid below.

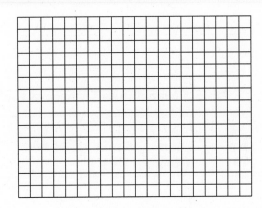

Wire diameter (mm)

4. Based on your graph, which kind of wire conducts energy more quickly—a thick wire or a thin wire?

_____

5. When roasting a large cut of meat, some cooks insert a metal skewer into the meat to make the inside cook more quickly. If you were roasting meat, would you insert a thick skewer or a thin skewer? Why?

_____

_____

_____

_____

Defending Your Conclusions

6. Suppose someone tells you that your conclusion is valid only for the particular metal you tested. How could you show that your conclusion is valid for other metals as well?

_____

_____

_____

_____

Copyright © by Holt, Rinehart and Winston. All rights reserved.

CHAPTER 9

9.7 **LABORATORY EXPERIMENT 9**

Determining the Better Insulator for Your Feet

(The lab corresponding to this datasheet begins on page 35 of Laboratory Experiments.)

Table 1 Mass of Warm Water Used to Make "Feet"

Bottle number	Mass of bottle (g)	Mass of bottle and water (g)	Mass of water (g)	Mass of water (kg)
1				
2				

Table 2 Temperature Changes of Each "Foot"

Bottle number/sock material	Initial temp. (°C)	After 4 min (°C)	After 8 min (°C)	After 12 min (°C)	Final temp. (°C)	Δt (°C)	Energy transferred by heat (J)
1/cotton							
2/wool							

Analyzing Your Results

1. Calculate the mass of water in each bottle in grams by using the following equation. Record your answers in **Table 1**.

   mass of water = mass of bottle and water − mass of bottle

2. Convert each mass of water to kilograms by using the following equation. Record your answers in **Table 1**.

   $$\text{mass of water (kg)} = \text{mass of water (g)} \times \frac{1 \text{ kg}}{1000 \text{ g}}$$

3. Calculate the change in temperature, Δt, for the water in each bottle by using the following equation. Record your answers in **Table 2**.

   $$\Delta t = \text{final temperature} - \text{initial temperature}$$

Copyright © by Holt, Rinehart and Winston. All rights reserved.

4. Calculate the energy transferred away from each "foot" by heat in joules by using the following equation. Use the value 4186 J/kg·K for the specific heat of water. Record your answers in **Table 2.**

$$\text{energy} = \text{specific heat} \times \text{mass of water (kg)} \times \Delta t$$

Reaching Conclusions

5. Which material, cotton or wool, allowed more energy to be transferred away from the "foot" by heat?

_____

6. How significant a difference is there in the insulating abilities of the two wet socks?

_____

_____

7. Which material do you recommend that hikers wear for socks when hiking in the cold rain?

_____

Defending Your Conclusions

8. In this experiment, you used a bottle of warm water to model energy transferred away from a hiker's foot by heat. How is this model different from the energy transferred away from a real foot by heat? How would you defend your results if someone claimed these differences made your results invalid?

_____

_____

_____

_____

_____

Expanding Your Knowledge

1. Using a similar model, design and perform an experiment that compares the insulating abilities of several different building materials.

2. Consult your local building materials store or library, or use the Internet to research the latest insulation developments in the construction or clothing industry. What kinds of materials are currently being developed and used?

3. Newer synthetic polyester clothing is advertised as having good insulating abilities in wet conditions. Test one or more of these synthetic materials against wool or cotton using your model.

4. Research the structure of wool to find out why it works so well in wet conditions.

Copyright © by Holt, Rinehart and Winston. All rights reserved.

How do particles move in a medium?

(The lab corresponding to this datasheet begins on page 328 of the textbook.)

Analysis

1. How would you describe the motion of the ribbon in step 2? How would you describe its motion in step 3?

_____

_____

_____

_____

_____

_____

2. How can you tell that energy is passing along the spring? Where does that energy come from?

_____

_____

_____

_____

_____

_____

_____

Copyright © by Holt, Rinehart and Winston. All rights reserved.

10.2 QUICK ACTIVITY, SECTION 10.1

Polarization

(The activity corresponding to this datasheet begins on page 329 of the textbook.)

Polarizing filters block all light except those waves that oscillate in a certain direction. Look through two polarizing filters at once. Then rotate one by 90° and look again. Explain what you observe in the space below.

_____

_____

_____

_____

_____

_____

_____

Copyright © by Holt, Rinehart and Winston. All rights reserved.

Wave Speed

(The activity corresponding to this datasheet begins on page 337 of the textbook.)

1. Place a rectangular pan on a level surface, and fill the pan with water to a depth of about 2 cm.

2. Cut a wooden dowel (3 cm in diameter or thicker) to a length slightly less than the width of the pan, and place the dowel in one end of the pan.

3. Move the dowel slowly back and forth, and observe the length of the wave generated.

4. Now roll the dowel back and forth faster (increased frequency). How does this faster motion affect the wavelength?

_____

_____

_____

_____

_____

5. Do the waves always travel the same speed in the pan?

_____

Copyright © by Holt, Rinehart and Winston. All rights reserved.

10.4 **DESIGN YOUR OWN LAB, CHAPTER 10**

Modeling Transverse Waves

(The lab corresponding to this datasheet begins on page 350 of the textbook.)

Designing Your Experiment

10. With your lab partner(s), decide how you will work together to make two additional sine curve traces, one with a different average wavelength than the first trace and one with a different average amplitude.

11. In the space below, write down your plan for changing these two factors. Before you carry out your experiment, your teacher must approve your plan.

_____

_____

_____

_____

_____

_____

_____

_____

Length along paper = 1 m	Time (s)	Average wavelength (m)	Average amplitude (m)
Curve 1			
Curve 2			
Curve 3			

Analyzing Your Results

1. For each of your three curves, calculate the average speed at which the paper was pulled by dividing the length of 1 m by the time measurement. This is equivalent to the speed of the wave that the curve models or represents.

CHAPTER 10

2. For each curve, use the wave speed equation to calculate average frequency.

$$\text{average frequency} = \frac{\text{average wave speed}}{\text{average wavelength}}$$

Defending Your Conclusions

3. What factor did you change to alter the average wavelength of the curve? Did your plan work? If so, did the wavelength increase or decrease?

_____

_____

_____

_____

_____

4. What factor did you change to alter the average amplitude? Did your plan work?

_____

_____

_____

_____

_____

Copyright © by Holt, Rinehart and Winston. All rights reserved.

Tuning a Musical Instrument

(The lab corresponding to this datasheet begins on page 39 of Laboratory Experiments.*)*

Table 1 Two Different Sounds Played Separately (Sounds 1 and 2)

Sound	Amplitude (V)	Time of first peak (s)	Time of last peak (s)	Elapsed time (s)	Number of waves between first and last peaks	Period (s)	Frequency (Hz)
1							
2							

Table 2 Two Sounds Out of Tune that Are Played Together (Sound 3)

Maximum amplitude (V)	Time of first maximum (s)	Time of last maximum (s)	Elapsed time (s)	Number of cycles between first and last maximums	Beat period (s)	Frequency (Hz)

Analyzing Your Results

1. Calculate the elapsed times for sounds 1 and 2 in seconds by using the following equation. Record your answers in **Table 1.**

$$\text{elapsed time} = \text{time of last peak} - \text{time of first peak}$$

2. Calculate the elapsed time for sound 3 by using the following equation. Record your answer in **Table 2.**

$$\text{elapsed time} = \text{time of last maximum} - \text{time of first maximum}$$

3. Calculate the periods for sounds 1 and 2 in seconds by using the following equation. Record your answers in **Table 1.**

$$period = \frac{elapsed\ time}{number\ of\ waves\ between\ first\ and\ last\ peaks}$$

4. Calculate the beat period for sound 3 by using the following equation. Record your answer in **Table 2.**

$$period = \frac{elapsed\ time}{number\ of\ cycles\ between\ first\ and\ last\ maximums}$$

5. Calculate the frequency in Hertz for all three sounds by using the following equation. Record your answers for sounds 1 and 2 in **Table 1** and your answer for sound 3 in **Table 2.**

$$frequency = \frac{1}{period}$$

Reaching Conclusions

6. Use your data for sounds 1 and 2 to describe the relationship between the pitch of a sound and frequency of the wave that is generated.

_____

7. Use the graph you printed or sketched for sound 3 to identify times when the two sound waves showed constructive interference. Why is the amplitude of the combined sound wave greater at these times than it is for each individual sound wave?

_____

_____

_____

Copyright © by Holt, Rinehart and Winston. All rights reserved.

Copyright © by Holt, Rinehart and Winston. All rights reserved.

8. Use the same graph to identify times when the two sound waves showed destructive interference. Why is the amplitude of the combined sound wave smaller at these times than it is for each individual sound wave?

9. Which method of tuning an instrument worked better for you, using the microphone and graphing calculator or simply listening for differences in pitches? Explain why.

Defending Your Conclusions

10. Musicians do not graph sound waves when they tune their instruments electronically. The electronic device they use tells them only whether the pitch is sharp (too high) or flat (too low). Does that make the results of your experiment invalid? Explain why or why not.

Expanding Your Knowledge

1. Interview a local piano tuner, and ask the tuner to demonstrate the equipment used to tune pianos.

CHAPTER 10

Charging Objects

(The activity corresponding to this datasheet begins on page 360 of the textbook.)

1. Rub two air-filled balloons vigorously on a piece of wool.

2. Hold your balloons near each other.

3. Now try to attach one balloon to the wall.

4. Turn on a faucet, and hold a balloon near the stream of tap water.

5. Explain what happens to the charges in the balloons, wool, water, and wall.

_____

_____

_____

_____

_____

_____

_____

Copyright © by Holt, Rinehart and Winston. All rights reserved.

11.2 QUICK ACTIVITY, Section 11.2

Using a Lemon as a Cell

(The activity corresponding to this datasheet begins on page 367 of the textbook.)

Because lemons are very acidic, their juice can act as an electrolyte. If various metals are inserted into a lemon to act as electrodes, the lemon can be used as an electrochemical cell.

1. Using a knife, make two parallel cuts 6 cm apart along the middle of a juicy lemon. Insert a copper strip into one of the cuts and a zinc strip the same size into the other.

2. Cut two equal lengths of insulated copper wire. Use wire cutters to remove the insulation from both ends of each wire. Connect one end of each wire to one of the terminals of a galvanometer.

3. Touch the free end of one wire to the copper strip in the lemon. Touch the free end of the other wire to the zinc strip. Record the galvanometer reading for the zinc-copper cell in your data table.

4. Replace the strips of copper and zinc with equally sized strips of different metals. Record the galvanometer readings for each pair of electrodes that you test in your data table. Which pair of electrodes resulted in the largest current?

Metals used in cell	Galvanometer reading (V)
Zinc-copper	

Copyright © by Holt, Rinehart and Winston. All rights reserved.

How can materials be classified by resistance?

(The lab corresponding to this datasheet begins on page 370 of the textbook.)

Material	Conducts electricity?
Glass stirring rod	
Iron nail	
Wooden dowel	
Copper wire	
Piece of chalk	
Strip of cardboard	
Plastic utensil	
Aluminum nail	
Brass key	
Strip of cork	

Analysis

1. What happens to the conductivity tester if a material is a good conductor?

_____

_____

2. Which materials were good conductors?

_____

_____

3. Which materials were poor conductors?

_____

_____

4. Explain the results in terms of resistance.

_____

_____

_____

Copyright © by Holt, Rinehart and Winston. All rights reserved.

11.4 QUICK ACTIVITY, Section 11.3

Series and Parallel Circuits

(The activity corresponding to this datasheet begins on page 375 of the textbook.)

1. Connect two flashlight bulbs, a battery, wires, and a switch so that both bulbs light up.

2. Is it a series or a parallel circuit? Make a diagram of your circuit in the space below.

_____

3. Now make the other type of circuit. Compare the brightness of the bulbs in the two types of circuits. Summarize your findings in the space below.

_____

_____

_____

_____

_____

_____

_____

_____

Copyright © by Holt, Rinehart and Winston. All rights reserved.

Test Your Knowledge of Magnetic Poles

(The activity corresponding to this datasheet begins on page 380 of the textbook.)

1. Tape the ends of a bar magnet so that its pole markings are covered.

2. Tie a piece of string to the center of the magnet and suspend it from a support stand.

3. Use another bar magnet to determine which pole of the hanging magnet is the north pole and which is the south pole. What happens when you bring one pole of your magnet near each end of the hanging magnet?

_____

_____

_____

_____

_____

4. Now try to identify the poles of the hanging magnet using the other pole of your magnet.

5. After you have decided the identity of each pole, remove the tape to check. Can you determine which are north poles and which are south poles if you cover the poles on both magnets?

_____

_____

_____

_____

_____

_____

Copyright © by Holt, Rinehart and Winston. All rights reserved.

11.6 QUICK ACTIVITY, SECTION 11.4

How can you make an electromagnet?

(The lab corresponding to this datasheet begins on page 384 of the textbook.)

Analysis

1. What type of device have you produced? Explain your answer.

_____

_____

_____

_____

_____

2. What happens to the direction of the compass needle after you reverse the direction of the current? Why does this happen?

_____

_____

_____

_____

_____

3. After detaching the coil from the cell, what can you do to make the nail nonmagnetic?

_____

_____

_____

_____

_____

Copyright © by Holt, Rinehart and Winston. All rights reserved.

11.7 SKILL BUILDER LAB, CHAPTER 11

Constructing Electric Circuits

(The lab corresponding to this datasheet begins on page 390 of the textbook.)

Description	Predicted value	Measured value
Current in circuit with a single resistor		
Current in circuit with two resistors in series		
Voltage across first resistor		
Voltage across second resistor		
Total current in circuit with two resistors in parallel		
Current in first resistor		
Current in second resistor		

1. If you have a circuit consisting of one battery and one resistor, what happens to the current if you double the resistance?

2. What happens to the current if you add a second, identical battery in series with the first battery?

3. What happens to the current if you add a second resistor in parallel with the first resistor?

4. Suppose you have a circuit consisting of one battery plus a 10 Ω resistor and a 5 Ω resistor in series. Which resistor will have the greater voltage across it?

5. Suppose you have a circuit consisting of one battery plus a 10 Ω resistor and a 5 Ω resistor in parallel. Which resistor will have more current in it?

6. Suppose someone tells you that you can make the battery in a circuit last longer by adding more resistors in parallel. Is that correct? Explain your reasoning.

Copyright © by Holt, Rinehart and Winston. All rights reserved.

Investigating How the Length of a Conductor Affects Resistance

(The lab corresponding to this datasheet begins on page 44 of Laboratory Experiments.)

Circuit Data

Distance between nails (cm)	Voltage (V)	Current (A)	Resistance (Ω)
2			
4			
8			
12			
16			
20			
24			
28			

Analyzing Your Results

1. Calculate the resistance of the circuit for each distance by using the following equation. Record your answers in your data table.

$$\text{resistance} = \frac{\text{voltage}}{\text{current}}$$

2. Plot your data in the form of a graph on the grid at right. Connect the data points with the line or smooth curve that fits the points best. If you use your graphing calculator, copy the graph from the calculator onto this datasheet.

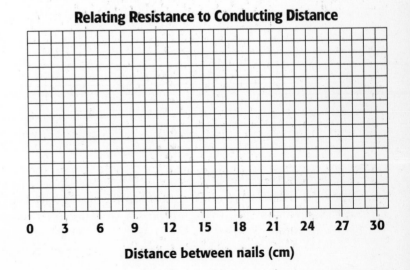

Relating Resistance to Conducting Distance

Resistance (Ω)

Distance between nails (cm)

Copyright © by Holt, Rinehart and Winston. All rights reserved.

Reaching Conclusions

3. What happens to the resistance of the circuit as the length of one of the conductors (the salt water) increases?

_____

_____

4. Both the salt water and the wires in the circuit are conductors. Which do you think conducts electricity more easily? Explain your answer.

_____

_____

_____

5. Every conductor has some resistance. Compare the resistance of salt water with that of the wires in the circuit.

_____

_____

_____

6. Suppose power is transmitted to your house from a transformer that is 45 m behind your house. The voltage supplied to your house is 110 V. If the power company guarantees a current of exactly 30 A at your breaker box (where the power enters your house), what is the maximum resistance per meter allowed in the power line to your house?

Defending Your Conclusions

7. How would your results be different if you had varied the length of one of the wires in the circuit instead of the conducting distance of the salt water?

_____

_____

_____

_____

Expanding Your Knowledge

1. Research the relative conductivity and cost of wires made from different metals. Summarize your findings in a poster that outlines the advantages and disadvantages of different types of wires.

Copyright © by Holt, Rinehart and Winston. All rights reserved.

12.1 QUICK ACTIVITY, SECTION 12.1
Examining Optical Fibers

(The activity corresponding to this datasheet begins on page 401 of the textbook.)

1. Carefully take apart a piece of fiber-optic cable.

2. In the space below, draw and label a diagram of what you find.

3. Point one end of a fiber toward a light source, and curve the fiber to the side.

4. Look at the other end. Does this show that light can move in a curved path? Explain.

_____

_____

_____

_____

_____

Copyright © by Holt, Rinehart and Winston. All rights reserved.

12.2 INQUIRY LAB, Section 12.2

How do red, blue, and green TV phosphors produce other colors?

(The lab corresponding to this datasheet begins on page 413 of the textbook.)

Analysis

1. Describe the three colors formed where two of the beams overlap.

_____

_____

_____

_____

_____

_____

2. What combinations of light produced the colors yellow and cyan?

_____

_____

_____

Copyright © by Holt, Rinehart and Winston. All rights reserved.

12.3 QUICK ACTIVITY, Section 12.3
How Fast Are Digital Computers?

(The activity corresponding to this datasheet begins on page 416 of the textbook.)

1. With a partner, use a stopwatch to time how long it takes for each of you to solve each set of the problems below (addition, subtraction, multiplication, and division) by hand. Record the times in the data table on the next page.

62 + 13	21 + 45	54 + 32	19 + 87	11 + 39
46 − 19	54 − 21	71 − 32	32 − 13	67 − 39
62 × 3	21 × 4	54 × 3	19 × 7	21 × 9
16)144	7)161	17)204	14)434	13)351

Now time how long it takes for each of you to solve the following problems using a calculator. (You do not need to copy the answers onto this datasheet.)

45 + 13	21 + 62	54 + 19	11 + 87	23 + 39
46 − 21	54 − 13	71 − 39	32 − 19	67 − 32
62 × 4	21 × 6	54 × 5	19 × 9	21 × 4
16)192	9)198	15)255	29)464	18)504

Copyright © by Holt, Rinehart and Winston. All rights reserved.

		Time (s)				
		Addition	**Subtraction**	**Multiplication**	**Division**	**Average**
You	**By hand**					
	By calculator					
Your partner	**By hand**					
	By calculator					
				Average time by hand		
				Average time by calculator		

2. Find the average time spent doing the problems by hand and by calculator. Record the two averages in the data table and compare them. Discuss your results in the space below.

_____

_____

_____

_____

_____

_____

_____

Copyright © by Holt, Rinehart and Winston. All rights reserved.

12.4 SKILL BUILDER LAB, CHAPTER 12

Determining the Speed of Sound

(The lab corresponding to this datasheet begins on page 426 of the textbook.)

Table 12-1 Data Needed to Determine the Speed of Sound in Air

	Tuning fork 1	Tuning fork 2
Vibration rate of fork (vps)		
Length of tube above water (mm), trial 1		
Length of tube above water (mm), trial 2		
Length of tube above water (mm), trial 3		

Analyzing Your Results

Table 12-2 Calculating the Speed of Sound

	Tuning fork 1	Tuning fork 2
Average measured length of air column (mm)		
Inside diameter (mm)		
Inside diameter × 0.4 (mm)		
Resonant length (mm)		
Wavelength of sound (mm), 4 × resonant length		
Speed of sound (m/s), wavelength × vibration rate of fork		
Speed of sound (m/s), calculated from item 4		

Copyright © by Holt, Rinehart and Winston. All rights reserved.

2. Measure the inside diameter of your tube and record this measurement in **Table 12-2.** The reflection of sound at the open end of a tube occurs at a point about four-tenths of its diameter above the end of the tube. Calculate this value and record it in **Table 12-2.** This distance is added to the length to get the resonant length. Record the resonant length in **Table 12-2.**

3. Complete the calculations shown in **Table 12-2.**

4. Measure the air temperature, and calculate the speed of sound using the equation shown below. Record your answer in **Table 12-2.**

> **speed of sound = 332 m/s at 0°C + 0.6 m/s for every degree above 0°C**

Defending Your Conclusions

5. Should the speed of sound determined with the two tuning forks be the same?

_____

6. How does the value for the speed of sound you calculated compare with the speed of sound you determined by measuring the air column?

_____

_____

_____

_____

_____

7. How could you determine the frequency of a tuning fork that had an unknown value?

_____

_____

_____

_____

_____

Copyright © by Holt, Rinehart and Winston. All rights reserved.

12.5 LABORATORY EXPERIMENT 12

Transmitting and Receiving a Message Using a Binary Code

(The lab corresponding to this datasheet begins on page 48 of Laboratory Experiments.*)*

Your Group's Binary Code

Symbol	Binary code	Symbol	Binary code

Analyzing Your Results

1. Interpret the graphs by writing the binary code for each symbol using 1's and 0's. Check with your lab partner to find out if your interpretation was right. Correct any errors you may have made.

_____

_____

2. What message did your lab partner send?

_____

_____

Copyright © by Holt, Rinehart and Winston. All rights reserved.

Reaching Conclusions

3. Compare the binary code you created with any other codes you have used. What are the advantages of using a binary code to send a message? What are the disadvantages?

_____

_____

_____

_____

_____

4. How could you change your code to include capital letters and punctuation marks?

_____

_____

5. The calculator can be programmed with a minimum sampling time of 0.000 164 s. How many symbols could you send in 1 minute using this sampling time? (**Hint:** First determine how many bits of information could be sent in 1 minute, and then divide this number by the number of bits you used to represent each symbol.)

6. How could an analog signal be sent using a light source? How would the graph of an analog signal differ from the ones you generated in this experiment?

_____

_____

_____

_____

Defending Your Conclusions

7. Intelligent life-forms in space would most likely not speak or understand the language your message was written in. Describe a message you could send that might be universally understood.

_____

_____

_____

Copyright © by Holt, Rinehart and Winston. All rights reserved.

Expanding Your Knowledge

1. Research the ASCII (American Standard Code of Information Interchange) code used in computers. In the space below, describe the code itself and how it is used in computers today.

_____

_____

_____

_____

_____

_____

_____

_____

2. There are several scientific research projects seeking intelligent life-forms in the universe with the use of radio telescopes. Create a poster that describes the goals, methods, and research activities of one of these projects. Two possible projects include the SETI (Search for Extraterrestrial Intelligence) and SERENDIP (Search for Extraterrestrial Radio Emissions from Nearby Developed Intelligent Populations).

Copyright © by Holt, Rinehart and Winston. All rights reserved.

CHAPTER 12

13.1 QUICK ACTIVITY, Section 13.1

Testing Foods for Fats

(The activity corresponding to this datasheet begins on page 435 of the textbook.)

1. Rub samples of five foods on squares of brown paper. Label each square with the name of the food that was rubbed on it.

2. After the squares have dried, compare the food stains that are left. In the space below, rank the five foods you tested in order of increasing fat content, based on the stain each food left on the paper.

_____

_____

_____

_____

_____

Copyright © by Holt, Rinehart and Winston. All rights reserved.

13.2 INQUIRY LAB, SECTION 13.1

How can you determine which foods contain the most vitamin C?

(The lab corresponding to this datasheet begins on page 437 of the textbook.)

Procedure

Type of juice	Drops of juice added to turn DCPIP clear
Lemon juice	

Analysis

1. Compare the juices you tested and the number of drops needed for each one. Does the addition of a greater number of juice drops indicate a higher vitamin C content or a lower vitamin C content?

_____

_____

_____

2. Of the juices you tested, which juice has the most vitamin C? Which juice has the least vitamin C? Did several juices have similar amounts of vitamin C?

_____

_____

_____

_____

3. What limitations could there be in using this test to compare the vitamin C content of different foods? (**Hint:** Would this test work well for a dark-colored grape juice?)

_____

_____

_____

_____

Copyright © by Holt, Rinehart and Winston. All rights reserved.

13.3 **QUICK ACTIVITY, Section 13.3**

Demonstrating the Importance of Mechanical Digestion

(The activity corresponding to this datasheet begins on page 450 of the textbook.)

1. Place 100 mL of water in each of the two beakers.

2. Measure the mass of a sugar cube.

3. Add the sugar cube to one of the beakers.

4. Take another sugar cube and crush it with a metal spatula or scoop. Pour these pieces into the second beaker. You can think of these pieces as a sugar cube that has been mechanically digested.

5. Observe how much sugar remains on the bottom of each beaker.

6. Did the sugar cube or the pieces of sugar dissolve more easily? Relate your findings to what happens when you digest food.

_____

_____

_____

_____

_____

_____

_____

_____

_____

_____

Copyright © by Holt, Rinehart and Winston. All rights reserved.

13.4 QUICK ACTIVITY, Section 13.3

Surface Area

(The activity corresponding to this datasheet begins on page 452 of the textbook.)

1. To model the difference between a large and a small surface area, you will need 64 small pieces of paper.

2. Place the pieces of paper on a table or on the floor.

3. Using one hand only, pick up as many squares as you can in exactly 10 s.

4. Repeat step 3 using two hands.

5. What would you expect if you had two people (four hands) picking up the squares?

_____

_____

_____

_____

_____

6. If the squares represent nutrients and your hands represent the intestinal wall, explain the advantage of greater surface area.

_____

_____

_____

_____

_____

_____

Copyright © by Holt, Rinehart and Winston. All rights reserved.

CHAPTER 13

13.5 SKILL BUILDER LAB, CHAPTER 13

Testing Food for Nutrients

(The lab corresponding to this datasheet begins on page 458 of the textbook.)

Tube	Contents	Nutrient tested	Observations	Result (+ or −)
1	Glucose solution	Sugar		
2	Unknown	Sugar		
3	Distilled water	Sugar		
4	Albumin solution	Protein		
5	Unknown	Protein		
6	Distilled water	Protein		

Analyzing Your Results

1. Based on your results, identify which nutrient(s) were present in the unknown sample.

2. Why was it important to test each indicator solution on known nutrient solutions and on distilled water?

Copyright © by Holt, Rinehart and Winston. All rights reserved.

Copyright © by Holt, Rinehart and Winston. All rights reserved.

Defending Your Conclusions

3. Suppose that the unknown sample did not produce a color change or a deposit when you added the indicator for one of the nutrients. Would that mean that the unknown sample was completely free of that nutrient? Explain.

_____

_____

_____

_____

_____

_____

CHAPTER 13

Analyzing Some Common Snack Foods

(The lab corresponding to this datasheet begins on page 52 of Laboratory Experiments.*)*

Table 1 Snack Food Data

	Peanut	Popcorn	Marshmallow	Corn tortilla chip
Mass of can (g)				
Mass of can and water (g)				
Food and stand mass before burning (g)				
Food and stand mass after burning (g)				
Temperature of water before burning (°C)				
Temperature of water after burning (°C)				

Table 2 Snack Food Results

	Peanut	Popcorn	Marshmallow	Corn tortilla chip
Water mass (g)				
Water mass (kg)				
Mass of food burned (g)				
Temperature change (°C)				
Energy content (Cal/g)				

Copyright © by Holt, Rinehart and Winston. All rights reserved.

Analyzing Your Results

1. Calculate the mass of water in grams that was in the can each time by using the following equation. Record your answers in **Table 2.**

$$\text{water mass} = \text{mass of can and water} - \text{mass of can}$$

2. Convert each mass of water to kilograms by using the following equation. Record your answers in **Table 2.**

$$\text{water mass (kg)} = \text{mass (g)} \times \frac{1 \text{ kg}}{1000 \text{ g}}$$

3. Calculate the mass of each food that burned in grams by using the following equation. Record your answers in **Table 2.**

$$\text{mass of food burned} = \text{food and stand mass before} - \text{food and stand mass after}$$

4. Calculate the temperature change of the water, Δt, each time by using the following equation. Record your answers in **Table 2.**

$$\Delta t = \text{temperature of water after} - \text{temperature of water before}$$

Copyright © by Holt, Rinehart and Winston. All rights reserved.

CHAPTER 13

5. Calculate the energy content of each food by using the following equation. Record your answers in **Table 2.**

$$\text{energy content} = \frac{\Delta t \times \text{water mass (kg)}}{\text{mass of food burned}} \times \frac{1 \text{ Cal}}{\text{kg} \cdot {}^{\circ}\text{C}}$$

Reaching Conclusions

6. Rank the snack foods you tested in order from the one with the greatest energy content to the one with the least.

7. Of the snack foods you tested, which one do you think is the most healthful? Explain why you think so.

Defending Your Conclusions

8. Your cells are able to "burn" nutrients and convert energy into heat and work. How is this process different in your cells than it is in this experiment? How is it the same?

Expanding Your Knowledge

1. Develop a procedure to test a variety of foods and compare their energy content.

2. Perform a similar experiment to test the energy content of several fuels. Try burning methanol, vegetable oil, paper, and other items to find out which one has the highest energy content.

Copyright © by Holt, Rinehart and Winston. All rights reserved.

14.1 INQUIRY LAB, SECTION 14.1

How does the structure of a heart relate to its function?

(The lab corresponding to this datasheet begins on page 468 of the textbook.)

Procedure

1. Place the mammalian heart in a dissecting pan.

2. Examine the heart closely, and compare it with the heart in **Figure 14-3** on page 466 of the textbook. Identify the heart's right and left sides. Locate the atria and the ventricles. Note your observations in the space below.

_____

_____

_____

_____

3. Locate the valves that control the flow of blood into and out of the ventricles. Note your observations in the space below.

_____

_____

_____

_____

4. Compare the thickness of the walls of the heart. Note your observations in the space below.

_____

_____

_____

Analysis

1. What explains the difference in wall thickness between the atria and the left and right ventricles?

_____

_____

_____

_____

14.2 **QUICK ACTIVITY, SECTION 14.1**

Direction of Blood Flow

(The activity corresponding to this datasheet begins on page 470 of the textbook.)

1. Find a large blood vessel that is visible on the inside of one of your forearms.

2. Using one finger, press firmly on the vessel near your elbow. Notice how the vessel swells *below* the point where your finger is compressing it.

3. As you continue to press on the vessel, slide your finger down your forearm. Notice that you can no longer see the vessel *above* the point where your finger is compressing it.

4. What do your observations in steps 2 and 3 indicate about the direction of blood flow in this vessel? What kind of vessel is it?

_____

_____

_____

_____

_____

_____

_____

Copyright © by Holt, Rinehart and Winston. All rights reserved.

14.3 **INQUIRY LAB, SECTION 14.2**

How much air can your lungs hold?

(The lab corresponding to this datasheet begins on page 475 of the textbook.)

Procedure

Balloon's circumference (cm)	

Analysis

1. Calculate the radius of the balloon by using the following formula.

$$\text{radius} = \frac{\text{circumference}}{2\pi}$$

2. Calculate the volume of the balloon (your vital capacity) by using the following formula.

$$\text{volume} = \left(\frac{4}{3}\right)\pi \times (\text{radius})^3$$

3. Compare your vital capacity with that of other students in the class. Is there much variation in vital capacity among students? If so, can you relate the variation to differences in weight, height, or physical fitness?

_____

_____

_____

_____

_____

_____

_____

Copyright © by Holt, Rinehart and Winston. All rights reserved.

CHAPTER 14

14.4 QUICK ACTIVITY, Section 14.2
Measuring Breathing Rate

(The activity corresponding to this datasheet begins on page 476 of the textbook.)

1. Sit quietly in your seat for about 5 minutes. Then measure how many times you breathe (an inhale accompanied by an exhale) in 30 s. Multiply this number by 2 to calculate your breathing rate in breaths per minute at rest.

2. Exercise vigorously by climbing stairs or performing some other strenuous activity for 5 minutes. As soon as you stop exercising, again measure how many times you breathe in 30 s. Multiply this number by 2 to calculate your breathing rate in breaths per minute after exercising.

3. Compare your breathing rate at rest with your rate after exercising. Did exercise produce any other change in your breathing besides your breathing rate?

_____

_____

_____

_____

_____

_____

_____

Copyright © by Holt, Rinehart and Winston. All rights reserved.

14.5 QUICK ACTIVITY, SECTION 14.3

Survey of Cigarette Smoking

(The activity corresponding to this datasheet begins on page 482 of the textbook.)

Do you smoke? (yes or no)	
If you do smoke, how many cigarettes do you smoke per day?	

Fold this paper in half so that your answers are not visible, and turn it in to your teacher. Your teacher will tabulate the results to find out what percentage of your class smokes cigarettes and the average number of cigarettes your class smokes per day.

Copyright © by Holt, Rinehart and Winston. All rights reserved.

14.6 DESIGN YOUR OWN LAB, CHAPTER 14

Studying the Effects of Exercise on Circulation and Respiration

(The lab corresponding to this datasheet begins on page 486 of the textbook.)

	Number of heartbeats in 15 s	Heart rate (beats/min)	Drops of NaOH solution added
Rest			
Exercise 1			
Exercise 2			
Exercise 3			
Exercise 4			

Designing Your Experiment

7. In the space below, form a hypothesis: How do you think exercise will affect heart rate and carbon dioxide production? A good way to exercise in the lab is to step onto and off of a step stool at a steady rate of one step every 2 s. How can you vary this procedure to change the intensity of exercise? Which measurements should you make?

_____

_____

_____

_____

In the space below, list each step you will perform in your experiment. Before you carry out your experiment, your teacher must approve your plan.

_____

_____

_____

_____

_____

_____

Copyright © by Holt, Rinehart and Winston. All rights reserved.

Analyzing Your Results

1. For each measurement of the number of heartbeats in 15 s, multiply by 4 to calculate heart rate in beats per minute. Record the results in your data table.

2.

Rest																								
Exercise 1																								
Exercise 2																								
Exercise 3																								
Exercise 4																								

Heart rate (beats/min)

Rest																								
Exercise 1																								
Exercise 2																								
Exercise 3																								
Exercise 4																								

Drops of NaOH solution added

3. How did your heart rate after exercise compare with your heart rate at rest?

_____

_____

4. How did the number of drops of sodium hydroxide solution you added after exercise compare with the number you added at rest?

_____

_____

Copyright © by Holt, Rinehart and Winston. All rights reserved.

5. Based on this experiment, what can you conclude about the effect of exercise on heart rate and carbon dioxide production?

_____

_____

_____

_____

6. What is the relationship between the intensity of exercise and changes in heart rate and carbon dioxide production?

_____

_____

_____

_____

Defending Your Conclusions

7. Someone repeated this experiment using a different type of exercise and found no effect on either heart rate or carbon dioxide production. Assuming the person tested correctly, list three reasons that might explain the results.

_____

_____

_____

_____

Copyright © by Holt, Rinehart and Winston. All rights reserved.

14.7 LABORATORY EXPERIMENT 14

Comparing the Respiration Rates
of Several Organisms

(The lab corresponding to this datasheet begins on page 56 of Laboratory Experiments.)

Table 1 Mass and Carbon Dioxide Data

Organism	
Mass of container (g)	
Mass of container and organism (g)	
Mass of organism (g)	
Concentration of CO_2 after 30 s (ppm)	
Concentration of CO_2 after 10 min (ppm)	

Table 2 Comparison of Respiration Rates

Organism	Concentration of CO_2 produced (ppm)	Rate of CO_2 production (ppm/s)	Respiration rate (ppm/s·g)
		Average	

Copyright © by Holt, Rinehart and Winston. All rights reserved.

Analyzing Your Results

1. Calculate the mass of the organism in grams by using the following equation. Record your answer in **Table 1.**

    mass of organism = mass of container and organism − mass of container

2. Calculate the concentration of CO_2 produced by the organism in parts per million by using the following equation. Record your answer in **Table 2.**

    CO_2 produced = concentration of CO_2 after 10 min − concentration of CO_2 after 30 s

3. Calculate the CO_2 production rate for the organism in parts per million per second by using the following equation. Record your answer in **Table 2.**

    $$\text{rate of } CO_2 \text{ production} = \frac{CO_2 \text{ produced}}{600 \text{ s}}$$

4. Calculate the respiration rate of the organism you tested in parts per million per second per gram by using the following equation. Record your answer in **Table 2.**

    $$\text{respiration rate} = \frac{\text{rate of } CO_2 \text{ production}}{\text{mass of organism}}$$

5. When you have finished with your calculations, write on the blackboard the respiration rate of the organism you tested. Record the respiration rates of all of the other organisms that were tested by your classmates in **Table 2.**

6. Calculate the average respiration rate for all of the organisms by adding the respiration rates and dividing by the number of organisms that were tested. Record the average in **Table 2.**

Copyright © by Holt, Rinehart and Winston. All rights reserved.

Reaching Conclusions

7. Rank the organisms that were tested from the one with the highest respiration rate to the one with the lowest respiration rate.

_____

_____

8. Use the observations you made and your knowledge of the organisms tested to rank them from the one that is most active to the one that is least active. Is there a relationship between an organism's activity level and its carbon dioxide production? What other factors may affect the respiration rates of different organisms?

_____

_____

_____

_____

_____

_____

_____

9. Suppose that the team that was studying plant gas exchange determined that the average plant consumes carbon dioxide at a rate of 3.6×10^{-3} ppm/s•g. What would be an appropriate ratio of animal mass to plant mass that would result in a balanced ecosystem? Show your calculation in the space below, and explain your answer.

_____

_____

_____

_____

_____

Copyright © by Holt, Rinehart and Winston. All rights reserved.

Defending Your Conclusions

10. Suppose someone tells you that your results are not valid because you didn't monitor the temperature in the container. How would performing the experiment at a higher temperature have affected your results? How would performing the experiment at a lower temperature have affected your results?

_____

_____

_____

_____

_____

_____

Expanding Your Knowledge

1. Set up a similar experiment to determine the average CO_2 consumption rate for plants. Then determine the ratio of plant mass to animal mass needed to balance an ecosystem.

Copyright © by Holt, Rinehart and Winston. All rights reserved.

DATASHEET

15.1 INQUIRY LAB, Section 15.1

How can you estimate the strength of a muscle?

(The lab corresponding to this datasheet begins on page 496 of the textbook.)

Estimated diameter of your biceps (cm)	

Analysis

1. Convert the diameter of your biceps to cross-sectional area by using the following equation. Record your answer in the table above.

$$\text{area (in cm}^2) = 3.14 \times \left(\frac{\text{diameter (in cm)}}{2} \right)^2$$

2. Now estimate the maximum strength of your biceps by using the following equation.

$$\text{strength (in N)} = \text{area (in cm}^2) \times 35 \text{ N/cm}^2$$

3. Is there a difference between your estimate and the amount of weight you know you can lift with that arm? (**Hint:** You can convert newtons to pounds by multiplying by a factor of 0.22.)

_____

_____

_____

4. Your forearm is like a lever, with the fulcrum at the elbow. Using what you learned in Chapter 8 about simple machines, can you offer at least one reason why the biceps' force you estimated (the input force) doesn't match the force you can actually lift (the output force)?

_____

_____

_____

_____

Copyright © by Holt, Rinehart and Winston. All rights reserved.

CHAPTER 15

15.2 QUICK ACTIVITY, Section 15.2

Determining Your Target Heart Rate

(The activity corresponding to this datasheet begins on page 504 of the textbook.)

Table 1 Heart Rate Data

Number of times your heart beats in 15 s	
Your resting heart rate (beats/min)	
Lower limit of your target heart rate (beats/min)	
Upper limit of your target heart rate (beats/min)	

1. Place your index finger and middle finger against the major artery on the underside of your wrist or on the side of your neck under your chin. Count the number of heartbeats that occur in 15 s and multiply this number by 4 to calculate your resting heart rate. Record both values in **Table 1.**

2. Your target heart rate is the rate at which your heart should beat when you are exercising. It should be within an acceptable range. This range is between 60 percent and 80 percent of your maximum heart rate. Using the table on page 504 in the textbook, calculate the lower and upper limits of your target heart rate. Record both limits in **Table 1.** Then compare your target heart rate with that of your resting heart rate.

_____

_____

_____

_____

_____

Copyright © by Holt, Rinehart and Winston. All rights reserved.

How can you measure your level of aerobic fitness?

(The lab corresponding to this datasheet begins on page 506 of the textbook.)

Table 1 Heartbeat Data

Number of heartbeats after resting for 1 min	
Number of heartbeats after resting for a total of 1.5 min	
Number of heartbeats after resting for a total of 2 min	
Total number of heartbeats	

Table 2 Assessing Your Fitness

Fitness level	Male	Female
Very fit	175 or less	190 or less
Fairly fit	about 200	about 220
Rather unfit	about 215	about 235
Very unfit	230 or more	250 or more

Analysis

1. Add the heartbeats you counted in steps 4–6, and record the total in **Table 1.** Compare this total with the numbers in **Table 2.**
2. In which fitness category did this test place you? Do you think this is a valid test of aerobic fitness? Explain why or why not.

_____

_____

_____

_____

_____

_____

_____

_____

Copyright © by Holt, Rinehart and Winston. All rights reserved.

15.4 QUICK ACTIVITY, Section 15.3

Simulating Weightlessness

(The activity corresponding to this datasheet begins on page 511 of the textbook.)

1. Place a stack of books on a desk, and move a wheeled swivel chair within arm's reach of the books. The desk and chair should be on a hard, smooth floor. Sit down in the chair, and lift your feet off the floor.

2. In this position, try to slide the stack of books along the desktop without touching anything but the books. First try to push the books directly away from you. Reposition the stack of books in front of you, and then try to push it to the right or to the left.

3. Using Newton's third law of motion, explain what happened in each case.

_____

_____

_____

_____

_____

_____

_____

_____

_____

_____

Copyright © by Holt, Rinehart and Winston. All rights reserved.

15.5 SKILL BUILDER LAB, CHAPTER 15

Comparing Skeletal Joints

(The lab corresponding to this datasheet begins on page 516 of the textbook.)

Observing Types of Joints

2. **Hinge Joint** Straighten one of your legs. Slowly bend the leg at the knee until your lower leg is folded behind you. Try to move the lower leg at the knee in other directions. Record your observations of the movements that are possible in the space below.

_____

_____

_____

_____

_____

_____

3. **Ball-and-Socket Joint** Move your upper arm in as many ways as possible from the shoulder. Record your observations of the movements in the space below.

_____

_____

_____

_____

_____

4. **Pivot Joint** Place your hands on both sides of your neck to hold your neck in place. Gently move only your head in all possible directions. Try not to let your neck bend. Record your observations of the movements that are possible in the space below.

_____

_____

_____

_____

_____

Copyright © by Holt, Rinehart and Winston. All rights reserved.

CHAPTER 15

5. **Saddle Joint** Move your thumb in as many ways as possible from its base. Record your observations of the movements that are possible in the space below.

_____

_____

_____

_____

_____

6. **Gliding Joint** Use your left hand to grip your right arm just above the wrist. Without moving your right forearm, move your right hand in all possible directions. Record your observations of the movements in the space below.

_____

_____

_____

_____

_____

7. **Semimovable Joints** Place the tips of your fingers on the lower part of your backbone. Move your torso in as many directions as possible. Record your observations of the movements in the space below.

_____

_____

_____

_____

_____

8. **Fixed Joints** Examine the joints in the top of a model of a skull. Gently press on each side of the joints to see if movement is possible without damaging the skull. Record your observations in the space below.

_____

_____

_____

_____

_____

Copyright © by Holt, Rinehart and Winston. All rights reserved.

Surveying the Joints in Your Body

9. Starting with your toes, examine the movement of each of the other joints in your skeleton. In the table below, record the other locations in the body where you discover each type of joint mentioned on the previous pages.

Joint type	Other locations of this joint in the body	Nonliving object with similar type of joint
Hinge joint		
Ball-and-socket joint		
Pivot joint		
Saddle joint		
Gliding joint		
Semimovable joints		
Fixed joints		

Analyzing Your Results

1. Rank the types of joints according to their ability to allow movement. Start with the joint that allows the least freedom of movement.

_____

_____

_____

_____

Defending Your Conclusions

2. Which types of joints are involved in walking?

_____

_____

_____

3. For each type of skeletal joint, record in the table above a common, nonliving object that has a similar type of joint in its construction.

Copyright © by Holt, Rinehart and Winston. All rights reserved.

15.6 **LABORATORY EXPERIMENT 15**

Developing a Physical Fitness Test

(The lab corresponding to this datasheet begins on page 60 of Laboratory Experiments.)

Table 1 Heart Rate Data

Activity		Test subject 1's heart rate (beats/min)	Test subject 2's heart rate (beats/min)
Sitting quietly for 2 min			
Standing quietly for 1 min			
After stepping for 30 s			
After stepping for 3 min			
Standing quietly for 5 min after stepping	After 30 s (0.5 min)		
	60 s (1 min)		
	90 s (1.5 min)		
	120 s (2 min)		
	150 s (2.5 min)		
	180 s (3 min)		
	210 s (3.5 min)		
	240 s (4 min)		
	270 s (4.5 min)		
	300 s (5 min)		

Copyright © by Holt, Rinehart and Winston. All rights reserved.

Table 2 Heart Rate Results

	Test subject 1	Test subject 2
Increase in heart rate after 30 s of exercise (beats/min)		
Maximum increase in heart rate (beats/min)		
Recovery time (s)		

Analyzing Your Results

1. Calculate your increase in heart rate at the start of exercise in beats per minute and that of your partner by using the following equation. Record your answers in **Table 2.**

increase in heart rate = heart rate after stepping for 30 s − standing heart rate

2. Calculate your maximum increase in heart rate in beats per minute and that of your partner by using the following equation. Record your answers in **Table 2.**

maximum increase = heart rate after stepping for 3 min − standing heart rate

3. Record your recovery time in seconds in **Table 2.** Recovery time is found by determining the number of seconds it took for your heart rate to return to within 5 beats per minute of your standing heart rate. Do the same for your lab partner. Remember that there are 60 s in 1 minute.

4. Copy your data onto a sheet of paper your teacher gives you. Do not put your name on the paper. Write down whether or not you regularly exercise. Give this paper to your teacher. When your teacher has averaged the data for your class, copy it from the board.

5. Plot your own heart rate data in the form of a graph on the grid on the next page that shows how your heart rate changed as you recovered from exercising. Also draw three horizontal lines on this grid to show your sitting heart rate, your standing heart rate, and your heart rate after 30 s (0.5 min) of exercising. Use a different color for each set of data, and be sure to label the graph clearly.

Copyright © by Holt, Rinehart and Winston. All rights reserved.

How Your Heart Rate Recovers After Exercising

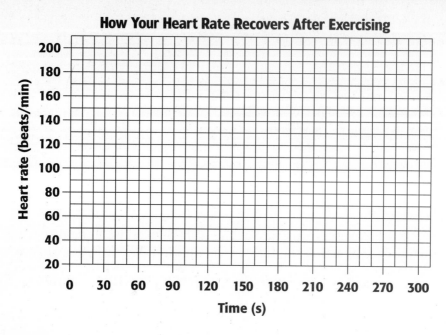

Reaching Conclusions

6. Use the data your teacher averaged to compare the increase in heart rate at the start of exercise, the maximum increase in heart rate, and the recovery times of people who exercise regularly with those of people who don't.

_____

_____

7. Do you think the step test will help camp leaders assess someone's overall fitness? Explain why or why not.

_____

_____

_____

Defending Your Conclusions

8. What factors besides fitness may affect someone's heart rate?

_____

_____

_____

Expanding Your Knowledge

1. Prepare an anonymous survey for your classmates to fill out asking about their daily routines, such as their diet and sleeping patterns. Also have each classmate include his or her heart rate data. Use a computer spreadsheet to analyze the completed surveys. Determine which factors are most closely related to someone's overall fitness.

Copyright © by Holt, Rinehart and Winston. All rights reserved.

Testing the Effect of Drugs on Heart Rate

(The lab corresponding to this datasheet begins on page 544 of the textbook.)

Solution	Heartbeats			Average number of heartbeats	Average heart rate (beats/min)
	Trial 1	**Trial 2**	**Trial 3**		
None					

Designing Your Experiment

7. With your lab partners, decide how you will test the unknown substances for their effects on daphnia heart rate. In which order should you test the different concentrations of each unknown substance? Should you use the same daphnia for all tests?

_____

_____

_____

_____

8. In the space below, list each step you will perform in your experiment.

_____

_____

_____

_____

_____

_____

_____

_____

9. Your teacher must approve your plan before you carry out your experiment.

Copyright © by Holt, Rinehart and Winston. All rights reserved.

Analyzing Your Results

1. For each row of data in your table, calculate the average number of heartbeats by adding the numbers for trials 1, 2, and 3 and dividing the sum by 3. Record the results in your data table.

2. Calculate the average heart rate in beats per minute by multiplying the average number of heartbeats by 6. Record the result in your data table.

3. Did unknown substance 1 increase or decrease the heart rate of daphnia? Did unknown substance 2 increase or decrease the heart rate? For each substance, which concentration produced the greatest change in heart rate?

_____

_____

_____

_____

_____

_____

4. What effect on heart rate would you expect a stimulant to have? What effect would you expect a sedative to have? Based on these expectations, classify each unknown substance as either a stimulant or a sedative.

_____

_____

_____

_____

_____

Copyright © by Holt, Rinehart and Winston. All rights reserved.

Defending Your Conclusions

5. Why was it important to use different daphnia for each unknown substance?

_____

_____

_____

_____

_____

_____

6. Is it reasonable to conclude that unknown substances 1 and 2 would have the same effects on heart rate in humans? Explain.

_____

_____

_____

_____

_____

_____

_____

_____

_____

Copyright © by Holt, Rinehart and Winston. All rights reserved.

16.2 LABORATORY EXPERIMENT 16

Investigating the Effects Detergents and Alcohols Have on Cells

(The lab corresponding to this datasheet begins on page 65 of Laboratory Experiments.)

Absorbance Data

Absorbance			
Deionized water	Ethanol	Rubbing alcohol	Dishwashing detergent
	5%:	5%:	0.05%:
	10%:	10%:	0.10%:
	20%:	20%:	0.20%:
	40%:	40%:	0.40%:

Analyzing Your Results

1. Plot your data for the ethanol and rubbing alcohol solutions in the form of a graph on the grid below. Plot both sets of data on the same set of axes, using a different color for each set of data. Be sure to label your graph clearly.

Comparing the Effects Ethanol and Rubbing Alcohol Have on Cells

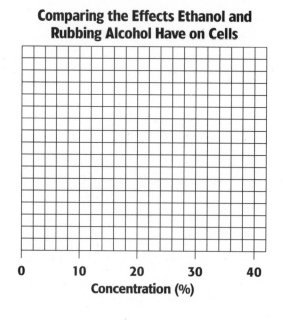

Absorbance

Concentration (%)

Copyright © by Holt, Rinehart and Winston. All rights reserved.

2. Plot your data for the detergent solutions in the form of a graph on the grid below.

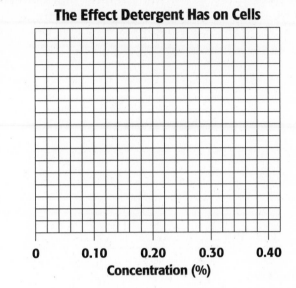

The Effect Detergent Has on Cells

Absorbance

Concentration (%)

0   0.10   0.20   0.30   0.40

Reaching Conclusions

3. Compare the effects ethanol and rubbing alcohol have on beet cell membranes. Which substance has a more damaging effect?

_____

_____

_____

4. Compare the effect dishwashing detergent has on cell membranes with that of ethanol. How does the detergent's effect compare with that of rubbing alcohol?

_____

_____

_____

_____

5. How does concentration affect the amount of damage done to cell membranes? Use your data to support your answer.

_____

_____

_____

_____

_____

Copyright © by Holt, Rinehart and Winston. All rights reserved.

Defending Your Conclusions

6. Suppose someone tells you that you should have plotted the data for all three substances on the same set of axes so that you could compare their effects more easily.
Explain why you plotted the detergent data on a separate graph. What would happen if
you plotted the data for all three substances on the same set of axes?

_____

_____

_____

_____

_____

_____

Expanding Your Knowledge

1. Repeat the experiment, testing a variety of household chemicals and cleaners. Group
the chemicals you test based on their ability to damage cell membranes. Summarize
the results of your experiment in the space below.

_____

_____

_____

_____

_____

_____

_____

_____

_____

Copyright © by Holt, Rinehart and Winston. All rights reserved.

17.1 QUICK ACTIVITY, SECTION 17.2

Thinking Twice

(The activity corresponding to this datasheet begins on page 558 of the textbook.)

The photographs on page 558 of the textbook show two sets of twins. One set is identical. Identical twins are often called maternal twins. The other set of twins is not identical—they are called fraternal twins.

1. How are fraternal twins produced? Explain your answer.

_____

_____

_____

_____

_____

_____

_____

_____

2. How are identical twins produced? Explain your answer.

_____

_____

_____

_____

_____

_____

_____

_____

Copyright © by Holt, Rinehart and Winston. All rights reserved.

CHAPTER 17

Investigating the Menstrual Cycle

(The lab corresponding to this datasheet begins on page 570 of the textbook.)

Graphing Hormone Levels

6. Use the information in the table on pages 570 and 571 of the textbook to plot the concentrations of FSH and LH each day in the form of a graph on the grid below. Use a different color for each hormone, and be sure to label your graph clearly.

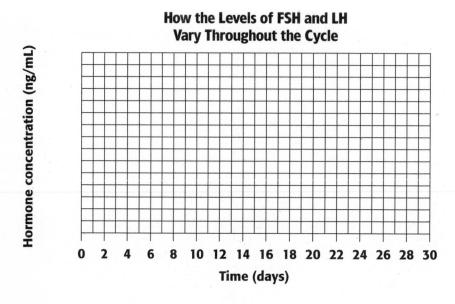

7. Use the information in the table on pages 570 and 571 of the textbook to plot the concentration of estrogen each day in the form of a graph on the grid below.

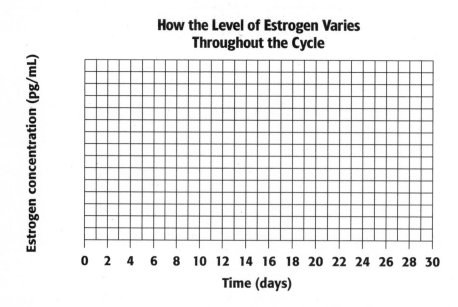

Copyright © by Holt, Rinehart and Winston. All rights reserved.

8. Use the information in the table on pages 570 and 571 of the textbook to plot the concentration of progesterone each day in the form of a graph on the grid below.

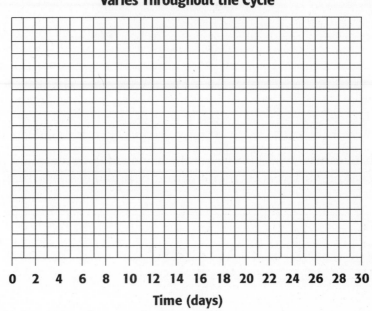

**How the Level of Progesterone
Varies Throughout the Cycle**

Progesterone concentration (ng/mL)

Time (days)

Analyzing Your Results

1. On what day does each hormone reach its maximum concentration?

_____

_____

_____

_____

_____

_____

2. Which hormones rise in concentration just before the midpoint of the cycle?

_____

_____

3. Which hormones rise in concentration gradually during the second half of the cycle?

_____

_____

Copyright © by Holt, Rinehart and Winston. All rights reserved.

Defending Your Conclusions

4. Ovulation is triggered by high concentrations of LH. How does the rise in LH concentration during the cycle correlate with the timing of ovulation?

_____

_____

_____

_____

_____

5. High levels of estrogen and progesterone are needed to maintain the thickened uterine lining. How do their levels correlate with the timing of menstruation?

_____

_____

_____

_____

_____

6. Progesterone limits the release of FSH and LH by the pituitary gland. How does the rise in progesterone concentration correlate with changes in the levels of FSH and LH?

_____

_____

_____

_____

_____

Copyright © by Holt, Rinehart and Winston. All rights reserved.

18.1 QUICK ACTIVITY, SECTION 18.1

Modeling the Universe

(The activity corresponding to this datasheet begins on page 582 of the textbook.)

1. Inflate a round balloon to about half full, then pinch it closed to keep the air inside.

2. Use a marker to draw several dots close together on the balloon. Mark one of the dots with an *M* to represent the Milky Way galaxy.

3. Now continue inflating the balloon. How do the dots move relative to each other?

_____

_____

_____

4. How is this a good model of the universe?

_____

_____

_____

_____

_____

5. In what ways might this not be a good model of the universe?

_____

_____

_____

_____

_____

18.2 QUICK ACTIVITY, SECTION 18.2

Using a Star Chart

(The activity corresponding to this datasheet begins on page 588 of the textbook.)

1. Locate the following stars on the star chart in Appendix B of your textbook: Betelgeuse, Rigel, Sirius, Capella, and Aldebaran.

2. What constellation is each star in?

Betelgeuse: _____

Rigel: _____

Sirius: _____

Capella: _____

Aldebaran: _____

3. Which of these stars appears closest in the sky to Polaris, the North Star?

_____

Copyright © by Holt, Rinehart and Winston. All rights reserved.

18.3 **SKILL BUILDER LAB, CHAPTER 18**

Estimating the Size and Power Output of the Sun

(The lab corresponding to this datasheet begins on page 604 of the textbook.)

Solar Viewer Data

Earth-sun distance, D (m)	1.5×10^{11}
Diameter of image, i (m)	
Pinhole-image distance, d_1 (m)	
Diameter of the sun, S (m)	

Analyzing Your Results

1. The ratio of the sun's actual diameter to its distance from Earth is the same as the ratio of the diameter of the sun's image to the distance from the pinhole to the image.

$$\frac{\text{diameter of sun, } S}{\text{Earth-sun distance, } D} = \frac{\text{diameter of image, } i}{\text{pinhole-image distance, } d_1}$$

Solving for the sun's diameter, S, gives the following equation.

$$S = \frac{D}{d_1} \times i$$

Substitute your measured values and $D = 1.5 \times 10^{11}$ m into this equation to calculate the value of S. Remember to convert all distance measurements to units of meters.

Solar Collector Data

Maximum temperature in sunlight (°C)	
Bulb-collector distance, d (m)	
Power of light bulb, b (W)	
Earth-sun distance, D (m)	1.5×10^{11}

2. The ratio of the power output of the sun to the sun's distance from Earth squared is the same as the ratio of the power output of the light bulb to the solar collector's distance from the bulb squared.

$$\frac{\textbf{power of sun, } P}{(\textbf{Earth-sun distance, } D)^2} = \frac{\textbf{power of light bulb, } b}{(\textbf{bulb-collector distance, } d)^2}$$

Solving for the sun's power output, P, gives the following equation.

$$P = \frac{D^2}{d^2} \times b$$

Substitute your measured distance for d, the known wattage of the bulb for b, and $D = 1.5 \times 10^{11}$ m into this equation to calculate the value of P. Remember to convert all distance measurements to units of meters. Your answer should be in watts.

Defending Your Conclusions

3. How does your value for S compare with the accepted diameter of the sun, 1.392×10^9 m?

_____

_____

_____

4. How does your value for P compare with the accepted power output of the sun, 3.83×10^{26} W?

_____

_____

_____

Copyright © by Holt, Rinehart and Winston. All rights reserved.

18.4 LABORATORY EXPERIMENT 18

Determining the Speed of an Orbiting Moon

(The lab corresponding to this datasheet begins on page 69 of Laboratory Experiments.*)*

Forces and Orbiting Speeds

Force (N)	Time to complete 20 revolutions (s)	Distance traveled in 20 revolutions (m)	Orbiting speed (m/s)
5.0			
7.5			
10.0			
12.5			
15.0			
17.5			
20.0			

Analyzing Your Results

1. Calculate the distance the mass travels in meters in one revolution by using the following equation. The radius is equal to 0.50 m.

$$\text{distance} = 2\pi \times \text{radius}$$

Multiply this distance by 20 to determine the total distance the mass traveled for each trial. Record your answer in your data table.

2. For each force, calculate the speed of the mass by using the following equation. Record your answers in your data table.

$$\text{speed} = \frac{\text{distance traveled in 20 revolutions}}{\text{time to complete 20 revolutions}}$$

Copyright © by Holt, Rinehart and Winston. All rights reserved.

3. Plot your data in the form of a graph on the grid below. Connect the data points with the line or smooth curve that fits the points best. If you are using your graphing calculator, copy the graph from the calculator onto this datasheet.

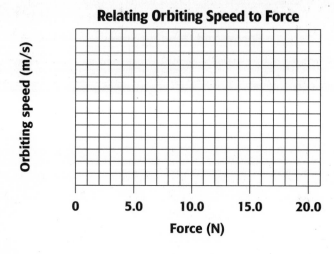

Relating Orbiting Speed to Force

Orbiting speed (m/s)

0   5.0   10.0   15.0   20.0

Force (N)

Reaching Conclusions

4. Use your graph to describe the relationship that exists between the orbiting speed of an object and the force acting on it.

_____

_____

_____

_____

5. Suppose that the opening animation of the film will show two planets. One of the planets will be twice as massive as the other planet. Compare the speeds of the moons orbiting these planets.

_____

_____

_____

_____

_____

Copyright © by Holt, Rinehart and Winston. All rights reserved.

6. You can determine the mass of a planet if you know the distance between the planet and one of its moons and the orbiting speed of the moon, as shown by the following equation.

$$\textbf{mass of planet} = \textbf{(speed)}^2 \times \textbf{radius} \times \left(1.499 \times 10^{10} \, \frac{\textbf{kg} \cdot \textbf{s}^2}{\textbf{m}^3} \right)$$

Suppose a planet and its moon are 250 000 000 m apart, and the speed of the orbiting moon is 320 m/s. Use the equation above to determine the mass of the planet in kilograms.

Defending Your Conclusions

7. Suppose someone tells you that your results are not valid because the masses and forces you measured in this experiment were too small to be comparable to a planet and its moon. How could you show that your results are valid?

_____

_____

_____

_____

_____

_____

Expanding Your Knowledge

1. Design an experiment similar to the one you just did to determine how changing the distance (the radius) would affect the orbiting speed of a moon. Perform the experiment, and summarize your findings in a report.

Copyright © by Holt, Rinehart and Winston. All rights reserved.

19.1 INQUIRY LAB, SECTION 19.1

Can you model tectonic plate boundaries with clay?

(The activity corresponding to this datasheet begins on page 615 of the textbook.)

Procedure

1. Use a ruler to draw two 10×20 cm rectangles on your paper, and cut them out.

2. Use a rolling pin to flatten two pieces of clay until they are each about 1 cm thick. Place a paper rectangle on each piece of clay. Using the plastic knife, trim each piece of clay along the edges to match the shape of the paper.

3. Flip the two clay rectangles so that the paper is at the bottom and place them side by side on a flat surface, as shown in the textbook on page 615. Slowly push the models toward each other until the edges of the clay make contact and begin to buckle and rise off the surface of the table.

4. Turn the models around so that the unbuckled edges are touching. Place one hand on each clay model. Slide one clay model toward you and the other model away from you. Apply only slight pressure toward the seam where the two pieces of clay touch.

Analysis

1. What type of plate boundary are you demonstrating with the model in step 3?

_____

2. What type of plate boundary are you demonstrating in step 4?

_____

3. How do the appearances of the facing edges of the models in the two processes compare? How do you think these processes might affect the appearance of Earth's surface?

_____

_____

_____

_____

_____

_____

_____

_____

Copyright © by Holt, Rinehart and Winston. All rights reserved.

19.2 SKILL BUILDER LAB, Chapter 19
Analyzing Seismic Waves

(The lab corresponding to this datasheet begins on page 642 of the textbook.)

Preparing for Your Experiment

2. P waves have an average speed of 6.1 km/s. S waves have an average speed of 4.1 km/s.

   a. How long does it take P waves to travel exactly 100 km?

   b. How long does it take S waves to travel exactly 100 km?

3. Because S waves travel more slowly than P waves, S waves will reach a seismograph after P waves arrive.

4. Use the time intervals found in step 2 to calculate the lag time you would expect from a seismograph located exactly 100 km from the epicenter of an earthquake.

City	Lag time (s)	Distance from city to epicenter (km)
Austin, TX		
Portland, OR		
Bismarck, ND		

6. The illustration on the top of page 643 of the textbook shows the records produced by seismographs in three cities following an earthquake.

Copyright © by Holt, Rinehart and Winston. All rights reserved.

7. Using the time scale at the bottom of the illustration, measure the lag time for each city. Be sure to measure from the start of the P wave to the start of the S wave. Ente your measurements in your data table.

8. Using the lag time you found in step 4 and the formula below, calculate the distance from each city to the epicenter of the earthquake. Enter your results in your table.

$$\text{distance} = \frac{\text{measured lag time}}{\text{lag time for 100 km}} \times 100 \text{ km}$$

Analyzing Your Results

1. Using the scale of the map below, adjust the drawing compass so that it will draw a circle whose radius equals the distance from the epicenter of the earthquake to Aust Then put the point of the compass on Austin, and draw a circle on your map. How i the location of the epicenter related to the circle?

_____

_____

_____

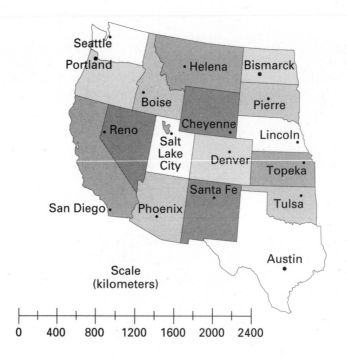

Copyright © by Holt, Rinehart and Winston. All rights reserved.

2. Repeat the process in item 1 using the distance from Portland to the epicenter. This time, put the point of the compass on Portland, and draw the circle. Where do the two circles intersect? The epicenter is one of these two sites.

_____

_____

3. Repeat the process once more for Bismarck, and find that city's distance from the epicenter. The epicenter is located at the site where all three circles intersect. What city is closest to that site?

_____

Defending Your Conclusions

4. Why is it necessary to use seismographs in three different locations to find the epicenter of an earthquake?

_____

_____

_____

5. Would it be possible to use this method for locating an earthquake's epicenter if earthquakes produced only one kind of seismic wave? Explain your answer.

_____

_____

_____

_____

6. Someone tells you that the best way to determine the epicenter is to find a seismograph where the P and S waves occur at the same time. What is wrong with this reasoning?

_____

_____

_____

_____

Copyright © by Holt, Rinehart and Winston. All rights reserved.

Relating Convection to the Movement of Tectonic Plates

(The lab corresponding to this datasheet begins on page 73 of Laboratory Experiments.)

Temperature Readings and Observations

Time (min)	Temperature (°C)				Observations
	Station 1	Station 2	Station 3	Station 4	
0					
0.5					
1					
3					
5					
10					

Analyzing Your Results

1. Plot your data in the form of a graph on the grid below. Plot the data for each station on the same set of axes, using a different color for each set of data. Be sure to label your graph clearly.

Comparing the Temperatures of the Stations

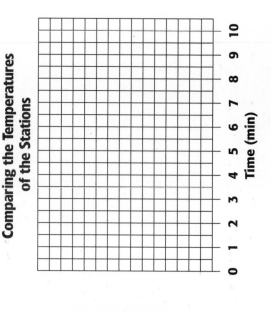

Temperature (°C)

Time (min)

Copyright © by Holt, Rinehart and Winston. All rights reserved.

2. Which station showed the greatest temperature change over the 10-minute period? Which station's temperature changed the least over the 10-minute period?

_____

_____

3. Use your data and your graph to describe how energy was transferred as heat between the stations in the aquarium.

_____

_____

_____

_____

Reaching Conclusions

4. Describe the rate at which food coloring left the jar when temperature differences were large. What happened to the rate as the temperature differences were smaller?

_____

_____

_____

5. Do your results support the theory that convection currents may be causing tectonic plates to move? Explain why or why not.

_____

_____

_____

_____

Copyright © by Holt, Rinehart and Winston. All rights reserved.

CHAPTER 19

6. If no new heat is being generated inside Earth, what does this experiment predict will eventually happen?

_____

_____

_____

_____

Defending Your Conclusions

7. To model convection currents in Earth's asthenosphere, you used an aquarium filled with cold water and a small jar filled with hot water. In what ways does Earth differ from your model? How could you change your model to make it more realistic?

_____

_____

_____

_____

Expanding Your Knowledge

1. Research how the solar system was formed. Include the sources of heat, the rates of cooling, and the substances that were present. Create a poster that outlines your findings.

Copyright © by Holt, Rinehart and Winston. All rights reserved.

20.1 QUICK ACTIVITY, Section 20.2

Measuring Rainfall

(The activity corresponding to this datasheet begins on page 659 of the textbook.)

1. Set an empty soup can outside in the open, away from any source of runoff. At the same time each day, use a metric ruler to measure the amount of rain or other precipitation that has accumulated in the can.

2. Record your measurements in the data table below. Keep a record of the precipitation in your area for a week.

3. Listen to or read local weather reports to see if your measurements are close to those given in the reports.

Day	Amount in can	Amount listed in weather report
1		
2		
3		
4		
5		
6		
7		

Copyright © by Holt, Rinehart and Winston. All rights reserved.

20.2 DESIGN YOUR OWN LAB, CHAPTER 20
Measuring Temperature Effects

(The lab corresponding to this datasheet begins on page 674 of the textbook.)

Temp. (°C)	Pull volume (mL)	Push volume (mL)	Average volume (mL)

Designing Your Experiment

7. With your lab partner(s), decide how you will use the materials available in the lab to determine the effect of temperature on air density. Test at least two temperatures below room temperature and two temperatures above room temperature. It is important that the mass of air inside the syringe does not change during your experiment. How can you ensure that the mass of air remains constant?

8. In the space below, list each step you will perform in your experiment.

_____

_____

_____

_____

_____

_____

_____

9. Before you carry out your experiment, your teacher must approve your plan.

Analyzing Your Results

1. At each temperature you tested, calculate the average volume by adding the pull volume and push volume and dividing the sum by 2. Record the result in your data table.

2. Plot your data in the form of a graph on the grid on the next page. Draw a straight line on the graph that fits the data points best.

Copyright © by Holt, Rinehart and Winston. All rights reserved.

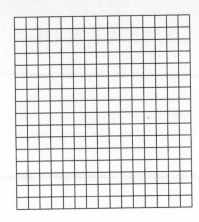

Temperature (°C)

3. How does the volume of a constant mass of air change as the temperature of the air increases? For the mass of air you used in your experiment, how much would the volume change if the temperature increased from 10°C to 60°C?

_____

_____

4. Recall that the density of a substance equals the substance's mass divided by its volume. Do your results indicate that the density of air increases or decreases as the temperature of the air increases? Explain.

_____

_____

_____

5. Based on your results, would a body of air tend to rise or sink as it becomes colder than the surrounding air? Explain.

_____

_____

_____

Defending Your Conclusions

6. Suppose someone tells you that your conclusions are invalid because some of your data points lie above or below the best-fit line you drew. How could you show that your conclusions are valid?

_____

_____

_____

Copyright © by Holt, Rinehart and Winston. All rights reserved.

20.3 LABORATORY EXPERIMENT 20
Predicting Coastal Winds

(The lab corresponding to this datasheet begins on page 77 of Laboratory Experiments.)

Temperature Changes of Water and Sand

Time (min)	Temperature of water (°C)	Temperature of sand (°C)
0		
1		
2		
3		
4		
5		
6		
7		
8		
9		
10		
Total temperature change		

Analyzing Your Results

1. Calculate the total temperature change for both the water and the sand by subtracting the initial temperature of each substance (temperature at 0 minutes) from the final temperature of each (temperature at 10 minutes). Record your answers in your data table.

Reaching Conclusions

2. Did the sand or the water have the greater temperature change during the 10-minute period?

_____

_____

Copyright © by Holt, Rinehart and Winston. All rights reserved.

3. Why was the temperature difference so large for this substance compared with the other substance?

_____

_____

_____

_____

4. Would you expect the winds to be blowing toward the shore or away from the shore in the middle of the day? Explain your reasoning.

_____

_____

_____

_____

_____

Defending Your Conclusions

5. Do you think your model of a coast is realistic? How does a real coastline differ from your model?

_____

_____

_____

_____

_____

_____

_____

Expanding Your Knowledge

1. Build and calibrate an instrument to measure the direction and speed of the wind at your school or at your home. Monitor the wind direction for several weeks, and try to develop a system for predicting winds.

Copyright © by Holt, Rinehart and Winston. All rights reserved.

21.1 INQUIRY LAB, SECTION 21.1
Why do seasons occur?

(The activity corresponding to this datasheet begins on page 685 of the textbook.)

Procedure

1. Place the lamp on a table, and turn the lamp on.

2. Stand about 2 m from the table, and hold the globe at arm's length, pointing it toward the lamp.

3. Tilt the globe slightly so that the bottom half—the Southern Hemisphere—is illuminated by the lamp.

4. Keeping the axis of the Earth's rotation pointing in the same direction, walk halfway around the table.

Analysis

1. What part of the globe is lit by the lamp's light now? What season does this represent in this part of Earth?

_____

_____

_____

_____

2. Would there be any seasonal changes if the Earth's axis were not tilted? Explain your answer.

_____

_____

_____

_____

3. In addition to experiencing seasonal changes, ecosystems also experience short-term changes as day changes into night. What movement of Earth causes night and day to occur?

_____

_____

_____

Copyright © by Holt, Rinehart and Winston. All rights reserved.

21.2 QUICK ACTIVITY, SECTION 21.3

The Effects of Acid Rain

(The activity corresponding to this datasheet begins on page 700 of the textbook.)

Rock	Initial mass (g)	Final mass (g)	Mass lost (g)	Percentage lost = mass lost/initial mass
Marble				
Limestone				
Granite				
Concrete				

1. Place a few small pieces of marble on a balance. Measure the initial mass, and record it in the table above.

2. Repeat step 1 with limestone, granite, and concrete.

3. Pour 50 mL of vinegar into each beaker of rocks. Observe and listen to the reactions in all four beakers. After 10 minutes, dispose of the vinegar as directed by your teacher.

4. Dry the marble and other rocks with a towel. Measure the final mass of the remaining rocks of each type, and record your results in the table above. Calculate the percentage of rock lost for each, and record the percentages in the table above.

Copyright © by Holt, Rinehart and Winston. All rights reserved.

21.3 QUICK ACTIVITY, SECTION 21.3

Observing Air Pollution

(The activity corresponding to this datasheet begins on page 701 of the textbook.)

1. Cut off a piece of masking tape about 8 cm long.
2. Place the sticky side of the tape against an outside wall, and press gently.
3. Remove the tape, and hold it against a sheet of white paper.
4. Did the tape pick up dust? If so, what might be the source of the dust?

_____

_____

_____

_____

5. Repeat the experiment on other walls in different places, and compare the amounts of dust observed.
6. Suggest reasons why some walls appear to have more dirt than others.

_____

_____

_____

_____

_____

Copyright © by Holt, Rinehart and Winston. All rights reserved.

How can oil spills be cleaned up?

(The lab corresponding to this datasheet begins on page 703 of the textbook.)

Procedure

1. Fill the pan halfway with cold water.

2. Pour a small amount of cooking oil into the water.

3. Try to clean up the "oil spill" using at least four different cleaning materials.

Analysis

1. Evaluate the effectiveness of each material. Which worked best? Explain why.

_____

_____

_____

_____

_____

2. Did any of the materials "pollute" the water with particles or residue? How might you clean up this pollution?

_____

_____

_____

_____

_____

_____

Copyright © by Holt, Rinehart and Winston. All rights reserved.

21.5 SKILL BUILDER LAB, CHAPTER 21

Changing the Form of a Fuel

(The lab corresponding to this datasheet begins on page 710 of the textbook.)

Data Table	
Mass of test tube A (g)	
Mass of test tube A with wood (g)	
Mass of wood (g)	
Mass of test tube A with solid residue (g)	
Mass of solid residue (g)	
Volume of gas, bottle 1 (mL)	
Volume of gas, bottle 2 (mL)	
Volume of gas produced (mL)	
Volume of test tube B (mL)	
Volume of liquid produced (mL)	

Analyzing Your Results

1. How much gas was produced? How much gas would be produced for 1 g of wood?

_____

_____

2. What happens when a burning splint is thrust into the gas?

_____

_____

3. Describe the contents of test tube B. How much liquid was produced for 1 g of wood?

_____

_____

4. What does the solid material remaining in test tube A look like?

_____

_____

Copyright © by Holt, Rinehart and Winston. All rights reserved.

5. How much solid material was left? How much solid material is that for 1 g of wood?

_____

_____

_____

6. Using insulated tongs, hold a piece of the solid material in the gas burner flame. How does it burn?

_____

_____

_____

Defending Your Conclusions

7. Why would you expect charcoal to give off little or no flame when it is burned?

_____

_____

_____

8. Why is this type of distillation called destructive?

_____

_____

_____

9. What do you think the liquids can be used for?

_____

_____

_____

_____

Copyright © by Holt, Rinehart and Winston. All rights reserved.

21.6 LABORATORY EXPERIMENT 21

Investigating the Effects of Acid Rain

(The lab corresponding to this datasheet begins on page 81 of Laboratory Experiments.*)*

pH Data

Drops of acid added	Deionized water	Buffer solution	Sample 1	Sample 2	Sample 3
0					
1					
2					
3					
4					
5					
6					
7					
8					
9					
10					

Copyright © by Holt, Rinehart and Winston. All rights reserved.

Copyright © by Holt, Rinehart and Winston. All rights reserved.

Analyzing Your Results

1. Plot your data in the form of a graph on the grid below. Plot each sample of water that you tested on the same set of axes, using a different color for each set of data. Be sure to label your graph clearly.

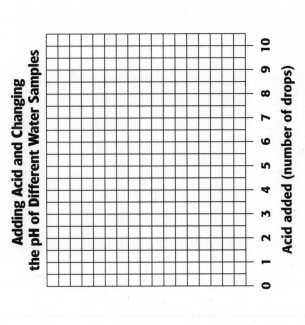

**Adding Acid and Changing
the pH of Different Water Samples**

pH

Acid added (number of drops)

0 1 2 3 4 5 6 7 8 9 10

2. Use your graph to compare the buffering capacities of deionized water and the buffer solution.

_____

_____

_____

_____

3. Which natural water sample demonstrated the most resistance to changing pH? Which sample demonstrated the least resistance? Use your data to support your answer.

_____

_____

_____

_____

Reaching Conclusions

4. Why does deionized water have such a low buffering capacity?

_____

_____

5. Explain why some of the water samples you tested have a greater buffering capacity than others.

_____

_____

_____

6. The pH scale is a logarithmic scale. This means that when pH decreases by one unit, the acid concentration increases by 10 times as much. Use this fact to explain why small changes in the pH of a body of water can be critical to the organisms that live there.

_____

_____

_____

Copyright © by Holt, Rinehart and Winston. All rights reserved.

Copyright © by Holt, Rinehart and Winston. All rights reserved.

7. Reactions within your body are constantly generating acids, such as lactic acid and carbonic acid. You may also eat many foods that are high in citric acid, acetic acid, and other acids. But the pH of your blood always stays between 7.35 and 7.45. Make a hypothesis about the buffering capacity of your blood. Which of the samples you tested is probably most like blood?

_____

_____

_____

_____

Defending Your Conclusions

8. Suppose someone suggests that one way to protect local lakes, streams, and rivers from acid rain is to add a buffer solution to the water. Do you think this is a good idea? Explain why or why not.

_____

_____

_____

_____

Expanding Your Knowledge

1. Test the buffering capacities of several different antacids using a process similar to the one you used in this experiment. Dissolve each antacid in water, then add acid one drop at a time while monitoring the pH. Determine which of the antacids you tested has the greatest buffering capacity.